ACS SYMPOSIUM SERIES **753**

Food Packaging

Testing Methods and Applications

Sara J. Risch, EDITOR
Science By Design

American Chemical Society, Washington, DC

Library of Congress Cataloging-in-Publication Data

Food packaging : testing methods and applications / Sara J. Risch, editor.

p.   cm.—(ACS symposium series , ISSN 0097–6156 ; 753)

Includes bibliographical references and index.

ISBN 0–8412–3617–8

1. Food—Packaging—Congresses.

I. Risch, Sara J., 1958– . II. Series.

TP374 .F656   2000
664ʹ.09—dc21   99–58687

The paper used in this publication meets the minimum requirements of American National Standard for Information Sciences—Permanence of Paper for Printed Library Materials, ANSI Z39.48–1984.

PRINTED IN THE UNITED STATES OF AMERICA

Foreword

THE ACS SYMPOSIUM SERIES was first published in 1974 to provide a mechanism for publishing symposia quickly in book form. The purpose of the series is to publish timely, comprehensive books developed from ACS sponsored symposia based on current scientific research. Occasionally, books are developed from symposia sponsored by other organizations when the topic is of keen interest to the chemistry audience.

Before agreeing to publish a book, the proposed table of contents is reviewed for appropriate and comprehensive coverage and for interest to the audience. Some papers may be excluded in order to better focus the book; others may be added to provide comprehensiveness. When appropriate, overview or introductory chapters are added. Drafts of chapters are peer-reviewed prior to final acceptance or rejection, and manuscripts are prepared in camera-ready format.

As a rule, only original research papers and original review papers are included in the volumes. Verbatim reproductions of previously published papers are not accepted.

ACS BOOKS DEPARTMENT

Contents

Preface

Packaging development has been driven by the demands of the food industry, regulatory agencies, and environmental groups. The food industry wants packaging materials that provide better barriers, that preserve products better, and that help give them longer shelf lives. The regulatory agencies want assurance that the materials, which are being developed, meet the necessary safety standards to insure that products are not contaminated by the materials in which they are packaged. On the environmental side, there are demands for use of less material overall and to reuse or recycle as much of that material as possible.

In the area of new barriers, not only are new materials being developed but also new ways of combining existing materials are being investigated. One approach that is being taken to enhance the quality of certain products is to use edible barriers within the food itself. An example of this might be a pizza in which the sauce could cause sogginess in the crust. A barrier between the two components could improve the quality of the final product. Another example where this technology might be applicable is in baked products containing pieces of fruit. The fruit has a higher moisture content than the surrounding baked good and a barrier could help to prevent the fruit from drying out and the baked product from getting soggy.

One of the areas of considerable interest that has been led by the environmentalists is the reuse and recycling of packaging materials. Both of these cause potential concern. Whether the material is to be recycled or reused, the concern centers on what the consumer may have used the package for after consuming the product and before returning it. Any contaminants that the package may have absorbed could potentially contaminate the next product that is put in the package. This could result in quality problems for a beverage that is put into a reusable bottle. If the bottle has absorbed a flavor, it may release it into the next product that is put into that container. If a nonfood item such as gasoline or a pesticide were stored in the bottle, there is potential for contamination of the food with unapproved chemicals.

With these developments come demands for better testing methodologies. When looking at barriers in packaging, there are standard tests for oxygen, water vapor, and other individual gases. These tests do not address the issues of aroma permeation or potential for migration from the packages. Extensive test development has occurred to prove the safety of both recycled and reused packaging materials. This testing has had to look at a wide variety of potential contaminants to insure that any possible scenario has been considered before the materials are approved for direct contact with food products.

Although testing can be carried out to determine the potential for migration from packaging materials into foods and the loss of flavors from the food through the packaging material, this is often a time consuming process. Research is being conducted to model these processes. Several chapters in this book address the issue of modeling the diffusion of compounds through different types of packaging. Numerous equations have been proposed and it appears that the modeling that is done is dependent on the compound that is diffusing through the material and the nature of the material itself. As these models are developed, they could provide a real benefit to package developers and regulatory agencies to better understand what might happen in actual use without having to conduct extensive testing of all the different potential migrants.

The packaging industry will continue to develop new materials, and scientists will be challenged to find ways to test those materials to prove their safety and effectiveness. This is the third American Chemical Society symposium to address the issue of interactions between foods and packaging materials. The needs of the marketplace will put demands on the materials that we use to better protect our foods and maintain not only the quality of the products but also the safety of the foods being consumed.

SARA J. RISCH
Science By Design
505 North Lake Shore Drive, #3209
Chicago, IL 60611

Chapter 1

New Developments in Packaging Materials

Sara J. Risch

Science By Design, 505 North Lake Shore Drive, #3209, Chicago, IL 60611

The development of new packaging materials, particularly those targeted for the food and agriculture industries, is a rapidly growing and changing arena. Packaging serves a variety of purposes for food products from simply protecting foods from outside contamination during distribution and storage to providing barriers that will maintain the correct moisture content, oxygen or carbon dioxide content in a product or maintain a desired atmosphere in the headspace around a product. This protection can be as simple as being a dust cover for the product to providing the means to maintain a modified atmosphere around a product for microbiological stability. In other cases the package can be designed to maintain the desired flavor profile of a product.

There are a number of key factors that are driving the growth and new developments in the chemistry of the packaging materials being used for food and other consumer products. One factor is the focus on the environment and the contribution of packaging materials to the solid waste stream. People are looking for ways to reduce the amount of packaging material that is being used, while still maintaining the integrity of the product. Another factor is the desire to have longer shelf life for products, which means developing better barrier properties in the packaging materials themselves. In many cases, manufacturers want to achieve source reduction and cost reduction while not compromising the product quality. This presents challenges to the packaging industry to develop thinner gauge and lighter weight materials that will perform the same function as the heavier weight counterparts. The globalization of the food industry is presenting challenges from the product quality and regulatory standpoint as materials to be exported must meet the requirements of whatever country into which they are to be shipped. All of these driving forces combined with the interest of packaging companies to develop new, interesting and value added items has resulted in a number of new developments.

Barrier Properties

One of the most challenging areas for packaging materials is that of barrier properties as this is readily measurable and has a tremendous impact on product quality. One area of research has simply been to combine materials through lamination, coextrusion, or coating that will give better barriers than the individual materials. Many materials are available today, in which the manufacturers have combined layers of different materials to give a complete package with the barrier properties desired. These layers can include foil, different types of plastic, paper, and adhesives. The plastic materials can include polyethylene (PE), polypropylene (PP), polyethylene terephthlate (PET), nylon, and ethylene vinyl alcohol (EVOH). In some cases this requires no new technology to laminate the materials together using. In others, compatibility of the materials may be any issue and development is required to cost effectively combine the desired materials.

As an example, oriented polypropylene (OPP) has relatively good moisture barrier properties but is a poor oxygen barrier when compared to a PET film. It is possible to combine these two materials in a lamination to achieve the benefits that both films can provide, but this will make a relatively thick film that is costly to produce. One development that has occurred is a film in which a barrier coating of amorphous polyester is coextruded onto the surface of biaxially oriented polypropylene (BOPP) (1). A development that has been reported in the area of combining materials to enhance the barrier properties is a patented plastic barrier layer that can be applied to paperboard which will give equivalent barrier properties to a foil lamination but is easier to recycle than a package made either with a thicker coating of plastic or with a layer of aluminum foil (2). It is claimed that this material can be used for packaging either dry or liquid products.

Metallization of packaging films is another technique that is used to enhance the barrier properties of plastic films, mainly PET and OPP. Metallized polyester (PET) has long been available with metallized OPP becoming more popular recently. As an example, the WVTR of a 1-mil sheet of OPP can be reduced from 0.4 to 0.1 and the OTR can be reduced from 150 to 3 by simply metallizing the film (3). It is important to treat the packaging material so that the metal layer will properly adhere and create the desired barrier. The treatments that are used are usually either patented or proprietary. One has been described as an ultra-high energy surface treatment (4). This treatment was used to produce a metallized film instead of plain film. The application of that film as the packaging material allowed one food manufacturer to extend the shelf of salty snack foods from either 6 or 9 months up to 12 months, allowing much broader distribution including shipments to international destinations. The end of the shelf life was defined as the time at which the snack foods would pick up enough moisture to become soggy and be rated as unacceptable from a texture standpoint.

Another coating that has received considerable attention but has achieved little commercial success is a coating of silicon dioxide. A thinly applied layer of SiO_2 does provide significant enhancement in barrier properties. The largest challenge with this coating has been to maintain the integrity of the coating during forming of a packaging material. The glass is susceptible to cracking when the film is run over a forming collar or the package is formed in another manner. An overview of glass-coated films was presented by Brody (5) in which he discussed the methods of coating a film, including sputtering, electron beam and plasma deposition. At that time in early 1994, the technology was still in its early stages of commercial development. Further work was reported by Finson and Hill (6) in which they had successfully printed and laminated films that the silicon oxide had been plasma coated onto films without losing the ultra-high barrier properties. As an example, the oxygen transmission rate (OTR) of uncoated PET was 115 $cc/m^2/day$ and the coated samples ranged from $1 - 2$ $cc/m^2/day$. The water vapor transmission rate (WVTR) of the same films decreased from 45 $g/m^2/day$ to between $1 - 5$ $g/m^2/day$. The testing indicated that the problem of cracking during conversion could be avoided and the barrier maintained if the film was coated using the proper method.

One new material that has been developed for use in packaging materials that has received considerable interest is polyethylene naphthalene (PEN). PEN is similar to PET except that naphthalene replaces terephthalate in the polymerization process. The resulting polymer has a rigid double ring in the backbone as opposed to the single ring in the backbone of PET. This creates a material with higher heat resistance, better moisture and gas barriers, better UV properties and overall generally improved strength (7). This product has been in development for a number of years and the use of the homopolymer finally received Food and Drug Administration (FDA) approval in April of 1996. An application for the use of PEN as a copolymer and in blends is still pending. While PEN does provide better barriers, the cost is significantly higher than PET. When approval is received, PEN could be used as a barrier layer, particularly for oxygen. This could provide for cost effective materials that use PEN as one layer instead of being the entire package structure. While this may help with the cost, it creates a material that may adversely affect the recycling of PET. It is difficult to separate the two polymers in the recycling stream.

One product category where manufacturers have been particularly interested is in the use of PEN is for beer bottles. There are many venues such as stadiums, beaches and other outdoor locations where the potential for breakage of glass limits or prevents its use. Plastic bottles are a good alternative; however, beer requires an excellent oxygen barrier to preserve the quality of the product. PEN as a homopolymer is being investigated and various combinations of materials are being used in an attempt to provide the barrier properties that are required to maintain the flavor of the beer.

There are plastic beer bottles on the market that are made only of PET which have limited distribution due to the fact that the shelf-life is only 6 weeks (8). One of these approaches has been to use a nylon layer sandwiched between two PET layers to provide the barrier needed (9). Another approach is a multi-layer bottle that incorporates a barrier layer of EVOH sandwiched between two layers of PET. While this bottle is being used, it only provides a shelf-life of 12 weeks (10) Unfortunately, some of these bottle compositions are proprietary and the exact materials are not being revealed. These multi-layer bottles do face challenges from those who are demanding that the materials be recyclable. To address this issue, one company has opted for PET that is sprayed with a layer of epoxy amine on the outside of the finished bottle. The coating increased the shelf life from 50 to 100 days, making it more viable for the retail trade (9). The bottle, being sold in Australia, meets their requirements for recyclability. The barrier coating can be removed and the bottle can then be recycled with other PET.

Another advantage of the coating is that colors can be incorporated into the coating to give an amber or green bottle, yet once the coating is removed, the PET to be recycled is clear.

One other material of interest in providing a barrier layer to packaging materials is liquid crystal polymers or LCPs. It should be noted that these materials are still in the experimental stage. They are called LCPs because in the molten state, the polymer molecules will mutually align and organize, as if they are in the crystalline state (11). LCPs can provide an oxygen transmission rate that is similar to ethylene vinyl alcohol (EVOH) as well as an excellent water vapor barrier. The largest drawback to the use of LCPs is their cost that will prevent them from being used on their own. The potential application is in multi-layer films where the LCPs can be used as a very thin layer to provide the barrier properties while other polymers such as PET and OPP can provide the structural integrity for the package.

High Permeability Films

A new product category that was virtually non-existent a few years ago has developed as a result of packaging materials that high levels of permeability to oxygen but still have low permeation rates for moisture. Fresh cut vegetables and salads have become a huge category in the supermarket. Once they are cut, the produce continues to respire, taking up oxygen and giving off carbon dioxide. Within the package, a modified atmosphere is created by this respiration (12). An increase in CO_2 levels up to about 8% will help to slow the respiration rate but higher levels can cause a decrease in the quality of the produce. The materials that have been developed for this market have varying OTRs, depending on the type of produce that is to be packaged. Broccoli has a high respiration rate and requires a package with an OTR of greater than 400 cc/100 in^2/day. Fresh cut salads need an OTR in the range from 100 to 400 cc/100 in^2/day (13).

Environmental Concerns

Packaging material has been the target of environmental and consumer activist groups as being a major contributor to the solid waste stream. The companies developing and using packaging materials have been working on ways to reduce the amount of packaging material being used, change the type of structure to be more environmentally friendly and develop materials that will be biodegradable.

One material that has shown promise as being biodegradable is polylactic acid, although it has yet to achieve any widespread use. One company that played a major role in the development of this material was Cargill, which has now set up a joint venture with Dow. Polylactic acid is a polymer made of repeating lactic acid units. Lactic acid exists as both d- and l-lactic acid. Controlling the amount of each during the polymerization process will affect the final properties of the material (14). Cargill Dow Polymers has scaled up their production and has found an application in yogurt cups being used in Germany (15). It is interesting to note that while corn is generally used as the source of material to produce the lactic acid to be polymerized, the anti-genetically modified sentiment in Germany has led them to use the sugar from sugar beets grown in

Spain that is fermented to produce lactic acid and shipped to the plant in the U.S. While the package may meet the requirements of being biodegradable, it is three times the price of the polystyrene or polypropylene that is generally used in this application.

Another biodegradable film that is in the early stages of development is a soy-based polyester film. A group at Michigan State University (16) has produced this material only on a laboratory scale. The potential applications are primarily for trash bags but the material could also be used for grocery sacks and as the film in bag in box cereal products. Soy protein is blended with polyester in amounts up to 40% soy protein and produces a material that can be injection molded or either blown or cast into a film. This material has not been commercialized but is the first time that soy protein has been successfully converted into a flexible material that does not become brittle and can be processed in conventional equipment. As with other new developments, cost is an issue with the price estimated to be at least twice that of low density polyethylene films that it would be likely to replace.

An example of an application of source reduction is in the production of agricultural chemicals. A package was developed using a lamination of nylon/foil/sealant is estimated to be two to three times stronger than the paper/paper/foil packaging that had been used and uses 43 percent less raw materials (17, 18). Spaulding also reported that a multi-layer package has been developed to package concentrates of lawn and garden chemicals. These materials are generally packaged in one-gallon rigid plastic containers can now be sold as concentrates and diluted by the end user. This results in a 97 % reduction in raw materials being used.

Other Developments

There are new developments in other areas of packaging as well. Many of the developments are held as trade secrets, making it difficult to evaluate the chemistry and technology that is being used. One area that has been under development for a number of years and has finally achieved commercial success is a new catalyst system. The system produces what is called a metallocene-based polyethylene (mPE) or other metallocene based polyolefins. This catalyst system has been instrumental in the development of films with high permeability that have been used for fresh produce, which was mentioned earlier. Metallocene based materials also have applications for materials to be used as sealant layers in multi-layer packaging. The catalyst offers films with good mechanical and optical properties (19).

Another area of interest is that of edible films. These are materials that can be used to protect food products or discrete components in a product and are made of materials that are approved as food materials. Research into edible films is covered later in this book.

One are that received considerable attention a few years ago is that of microwave packaging. The one key development that has been around for a number of years is that of the microwave susceptor, which is a metallized film that interacts with microwave energy to heat up in the oven. This film is typically PET that has aluminum vacuum deposited on it. This can be incorporated into a package by either laminating it between two sheets of paper to produce a flexible package or by laminating it to paperboard to produce a rigid container. The most common use of the flexible package is for microwave popcorn, while the rigid container has been used for pizza, fish and sandwiches. One of the potential drawbacks of the susceptor is that it does heat up to approximately 400 F and the packaging material may scorch if there is not a food product in direct contact with it. One place where this was potentially a problem was with very low or no added fat microwave popcorn. Regular microwave popcorn has added oil in contact with the susceptor to absorb the energy. The lower fat varieties do not have the oil to serve as a heat sync. One material that has been developed to address this issue is a susceptor with tiny X's etched into the aluminum layer. This causes the susceptor, called the Safety Susceptor (20) to quit heating when it reaches a certain temperature. The use of pattern susceptors has also been promoted to give targeted heating in a product (21).

There have also been developments in passive packaging that simply holds a product during cooking. One of the most common examples of a cost effective and efficient package is PET coated paperboard. This package is durable and can go directly from the freezer into the microwave. Many frozen, microwavable entrees today use CPET (crystalline polyester). A new development in this area has been to inject carbon dioxide or nitrogen into the resin to make a foamed CPET. This material does not have the strength of barrier properties of the regular CPET but can be used for refrigerated foods that have been designed to be reheated in the microwave.

While there have been some new ideas recently, there have not been any true breakthroughs that have had commercial significance. After numerous new product introductions during the late 1980's there was significantly decreased interest in new microwavable foods and packaging materials. One of the biggest issues was that the packaging materials that provided the desired performance were multiples of the cost of the non-microwavable counterpart. Many of the materials that were in use 10 to 15 years ago are still being used successfully today. Some work is continuing to confirm the safety of the materials that do reach higher temperatures when used in the microwave, however, it is not a major issue as it was in the early 1990's. Several chapters later in this book address the research that is being conducted to confirm the safety of not only microwavable packaging materials but other materials as well.

The packaging industry will continue to develop new materials to meet the challenging needs of today's food products. In some cases these materials will be lighter weight yet provide better barriers to moisture, oxygen, other gases and aromas. These materials may be combinations of existing polymers or may be newly developed materials. In other cases, the packages will be designed to have the specific permeability that is needed to maintain the desired atmosphere around a "living" product such as is seen with the packaging materials designed for fresh produce. Packaging will continue to move from simply being a container that covers the product to a material that plays an active role in the quality of the food. While the development of new microwavable packaging materials has slowed down, there is still interest in this type of active packaging and companies are continuing to develop packages that will enhance the cooking performance in the microwave oven. Packaging companies will work with food companies to better understand how products fail or reach the end of their shelf life so that the materials being developed will meet the demands of the marketplace. Packaging plays an integral role in the quality and safety of our food supply and the developments will continue to insure a higher quality and safer food supply than we have today.

Literature Cited

1. Mueller, T. R. *Proceedings of the TAPPI Polymer, Coatings and Lamination Conference,* Technical Association of the Paper and Pulp Industry, Atlanta, GA, 1997.
2. Anon. *Dairy World*, 1998, *181*, 53.
3. Stauffer, C.E. *Baking and Snack*, *18*, 54.
4. Anon. *Packaging World*, 1998, *5*, 2.
5. Brody, A.L. *Pack. Tech. Eng.* 1994, *3*, 44.
6. Finson, E. and Hill, R. J. *Pack. Tech. Eng.* 1995, *4*, 36.
7. Morse, P.M. *C&E News*, 1997, *75*, 8.
8. Tilley, K. *Plastics News*, 1997, *9*, 24.
9. Reynolds, P. *Packaging World*, 1998, *5*, 38.
10. Reynolds, P. *Packaging World*, 1998. *5*, 54.
11. Lusignea, R.W. *Pack. Tech. Eng.* 1997, *6*, 38.
12. Russell, M. *Food Eng.* 1996, *68*, 37.
13. Hoch, G.J. *Food Proc.* 1998, *59*, 29.
14. Ryan, C.M., Hartmann, M.H., Nangeroni, J.F. *Pack. Tech. Eng.* 1998, *7*, 39.
15. Anon. *Packaging World*, 1998, *5*, 2.
16. Newcorn, D. *Packaging World*, 1998, *5*, 94.
17. Spaulding, M. *Converting Mag.* 1998, *16*, 47.
18. Merola, M. *Packaging World*. 1998, *5*, 58.
19. Strupinski, G. *Pack. Tech. Eng.* 1997, *6*, 20.
20. U.S.patent 5,412,187.
21. Demetrakakes, P. *Food Processing* 1997, *58*, 94.

Chapter 2

Edible Barriers: A Solution to Control Water Migration in Foods

Frédéric Debeaufort[1,2], Jesús-Alberto Quezada-Gallo[1,3], and Andrée Voilley[1]

[1]ENS.BANA, Laboratoire de Génie des Procédés Alimentaires et Biotechnologiques, Université de Bourgogne, 1 Esplanade Erasme, 21000 Dijon, France
[2]I.U.T. Génie Biologique, Boulevard du Dr. Petitjean, BP 510, 21014 Dijon Cedex, France
[3]CONACyT, Mexico

The loss of food quality depends often on migration of small molecules such as water, salts, pigments or aroma compounds. traditional packagings allows to reduce transfer, but only between food and the surrounding medium. Edible films and particularly coatings allows it too, but they can be applied inside the food such as between fruits and baked pastry in a pie. A wide range of substances from animal or vegetal origin can be used to formulate an edible barrier. The formulation have to be set up as a function of both food composition and hedonic property, and of the nature of the migrant. Thermodynamics and kinetics of the transfer mechanism, the structure of the barrier and the nature of the diffusing substance, affect tremendously the barrier performances of edible packagings.

One of the most important problems occurring in food preservation is mass transfer between the product and its surrounding medium and/or between two different parts in a product. An ideal food package should control such mass transfer and provide mechanical protection against the mechanical stress during production, transport and storage. Controlling mass transfer requires an understanding of the role of the interface between two parts on mass transfer rates.

The consequences of mass transfer in food are often modifications of food texture, color, flavor and aroma during storage, but mass transfer also may induce physico-chemical, biochemical and microbiological alterations, leading to a loss of quality, *i.e.* the loss of crispness of a cereal product, oxidation of polyunsaturated lipids in meats or mould growth in dried fruits (*1, 2, 3*).

Mechanism of Water Transfer

To better understand the mechanism of the water transfer in food, we need to understand the relationship between water content and water activity in the food. An example of migration between two parts in a product can be the water migration between biscuit dough and raisins, as described by Karathanos and Saravacos (4).The water content of dough was initially higher than that of raisins, the water activities too. After three days contact between dough and raisins, the water content and the water activity in the dough decreased whereas they increased in the raisins. At equilibrium, the water content of raisins was higher than in dough, but the activities were the same. This work confirmed the theory of transfer which depends only on the chemical potential gradient. Thus, moisture migration is governed only by the water activity (or the partial pressure or the molar fraction), and not by water content, as often considered in food industry.

Furthermore the phenomenon of water transfer induces swelling and bursting of raisins, growth of molds, drying of the dough and some troubles in process. In this case, there is no contribution of the traditional packaging to limit this problem. Indeed we cannot imagine applying a plastic film for wrapping raisins as a moisture barrier.

At equilibrium (infinite time), water activities of the two compartments tend to reach the same value, while the water contents will not if the sorption isotherms of the two compartments are different (Figure 1). So, water transfer *is only controlled by activity gradients*. This observations suggest two means of preventing (or minimizing) water transfer in foodstuffs :

a) The closer the water activities of the different food components, the lower the migration will be. This means a change in the formulation of the food. However, it works well only for small water activity gradients.

b) In all the cases, a barrier layer can be applied at the interface of the two parts of the product or between the product and the surrounding medium. These barrier layers are commonly called edible packaging and considered as a food component.

Edible Packagings and Coatings

When a packaging like a film, a sheet, a thin layer or a coating is an integral part of the food and/or can be eaten with, then it is qualified as edible packaging. This packaging can be a film or a sachet if its structure is independent of the product, and a coating if it is a food integrated structure. Taking into account the problems to be solved by edible barriers, an edible packaging must have the followings properties (5,6).

Firstly, it must be edible, depending on the laws of the country where the product will be consumed. It must be free of toxic compounds and must have a high

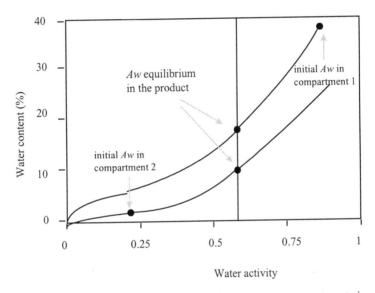

Figure 1. Mechanism of water equilibration between two compartments in an heterogeneous food.

biochemical, physico-chemical and microbiological stability, before, during and after application. Edible packaging should have good sensory qualities (or at least be tasteless), and have good barrier and mechanical efficiencies. Moreover, if they are non-polluting, perfectly biodegradable, have a simple technology and low cost, it is much better.

An edible packaging is composed of at least two substances :
   a) a film forming substance giving cohesiveness to a continuous matrix
   b) and a barrier compound providing impermeability to the film or coating.

Main components of edible packaging are food additives and typical food ingredients, *i.e.* polysaccharides, proteins and/or lipids. Some additives are often used to improve functional properties of films and coatings :

Plasticizers, such as polyols or fatty acids or small molecules, increase the mechanical deformation of films, making handling easier. *Emulsifiers,* like mono- and di-glycerides or lecithins, can be used to reduce the fat globule size and to increase their distribution in emulsified edible films, increasing at the same time the overall hydrophobicity of the barrier. *Acids and alkalis* are used to improve the solubilization of the biopolymers, mainly in the case of proteins and the homogeneity of the network and thus the mechanical resistance of the packaging. *Surface active substances* improve adhesiveness of edible coatings on supports.

The application of edible films or coatings on the surface of a food or at the interface between two parts within a same food can limit the transfer of compounds such as: water, organic vapors (aroma compounds or solvents), some gases (oxygen, carbon dioxide, nitrogen and methane), or several non volatile solutes (lipids, salts, pigments or additives). Edible coatings could also provide a barrier against light or UV which can modify the food characteristics via oxidation of lipids and pigments (Figure 2). However, there is no value to applying a very efficient barrier if it is very brittle since mass transfer may occur across cracks. Therefore, edible packagings have to withstand mechanical stresses.

A comparison of the efficiency of edible packagings with that of plastic packagings (Tables I and II), shows that the mechanical characteristics of polysaccharides and proteins are of the same order of magnitude as cellophane and low density polyethylene (LDPE). However, the permeability of waxes to water transfer can be 20 times lower than a hydrophobic polymer like LDPE. The aroma compound and gas barrier properties may also be very similar. There are exception such as edible films based on wheat gluten and glycerol. These films are 500 times less permeable to oxygen than polyethylene and 10 times less permeable to 1-octen-3-ol, a mushroom-like aroma compound.

Edible Packagings: Realities

The main quality required for edible films is to control mass transfer. In most cases, edible packagings are used to limit water migration. In this aim, Kamper and Fennema (7) applied a composite film based on hydroxypropyl methylcellulose and beeswax to retard the balance of water activities between two food components of

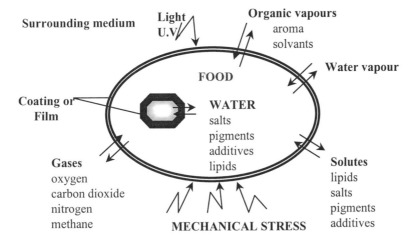

Figure 2. Required properties of edible packagings.

Table I. Comparison of the Barrier Efficienties of Edible and Plastic Packaging

Permeability (g.m⁻¹.s⁻¹.Pa⁻¹)

FILMS		Water ($\times 10^{11}$)	Oxygen ($\times 10^{15}$)	1-octen-3-ol ($\times 10^{11}$)	2-heptanone ($\times 10^{11}$)
Edible	Methylcellulose	7.0	2.45	12.20	3.9
	Wheat gluten + glycerol	9.2	0.06	0.46	n.d
	Edible wax	0.02	15.36	n.d	n.d
Synthetic	Cellophane	5.6	0.26	0.15	< 0,1
	LDPE	0.1	30.60	5.10	23,9

Table II. Comparison of the Mechanical Efficienties of Edible and Plastic Packaging

Mechanical resistance

FILMS		Tensile Strength (MPa)	Elongation (%)
Edible	Methylcellulose	41	40-180
	Wheat gluten + glycerol	5	20-90
	Methylcellulose + Triglycerides	28	20-100
Synthetic	Cellophane	77	30
	LDPE	9	200

fresh or frozen products (pizza, stuffed biscuits). The film applied showed good barrier properties against water transfer, but it had an adverse effect on sensory quality of the product. The same authors (8) used a methylcellulose and fatty acid-based film on the same frozen food, and completed the physico-chemical study with a sensorial analysis. In this case, the sensory quality of the product remained acceptable after 10 weeks storage at -6.5°C , while they were lost after 3 weeks when no barrier coating were applied.

To delay water absorption which results in the loss of crispness in dry biscuits after the plastic packaging is taken off, a gluten-margarine-based coating has been applied in our laboratory with success. The shelf-life of the dry cereal product increased by at least 30% under severe conditions of storage (100% R.H. at 25°C).

Nussinovitch and Hershko (9) applied a carrageenan-based coating on cloves of garlic with the aim of limiting their dehydration during carriage and storage. An interesting result was that the water loss in the product with a coating was more regular and slower especially in the first days after coating.

Edible coatings allows to control the water absorption and the retention of a preservative agent at the surface of dried fruits, for example to inhibit mold growth (3). Moreover, the application of edible coatings can control gas exchanges (O_2 and CO_2) and significantly prevent rancidity of polyunsaturated fats. Indeed, a whey protein and monoglyceride-based coating reduced water loss and lipid oxidation during storage of frozen salmon slices (2).

Additionally, edible packaging can be applied to retain aroma compounds within a food. Sensidoni et al. (10) used several composite coatings on a model liquor used in chocolate toffee. At least 60% of the aroma compounds of the stuffing liquor were retained in the toffee. The effect of visible and UV lights inducing e.g. reactions in foods could be reduced, particularly in meat as showed by Peyron (12).

In summary, proteins and polysaccharides exhibit good mechanical and sensory properties and they represent a very good barrier against gas and aroma compound transfers, whereas lipids are the best moisture edible barrier. That is the reason why, in most cases, composite films including at least a biopolymer (protein or polysaccharide) and a lipid are applied. Edible packaging has potential applications when plastic packagings can not be used. Thus it may be worthwhile to take advantage of both edible and plastics packagings to preserve food quality.

Literature Cited

1. Debeaufort, F.; Voilley, A.. Fonctions et applications des emballages comestibles. Colloque « l'Aliment pour Demain : Au Delà de la Fonction Nutritionnelle », 25-26 Janvier, **1996,** Dijon.
2. Stuchell, Y.M.; Krochta; J.M.. *J. Food Sci.*, **1994**. *60*, , 28-31.

3. Guilbert, S. Gontard, N.. Application de substances filmogènes aux fruits et légumes. In *Formation Inter-entreprises ADRIA. Substances alimentaires filmogènes et leurs applications.*. Paris. **1995**.

4. Karathanos V. T. and Kostaropoulos A. E.. *J. Food Eng.* **1995**, *25*, 113-121.

5. Gontard; N.; Guilbert; S. In *Food packaging and preservation*, Mathlouthi, M., Ed.; Blackie Academic and Professional, London, **1994**, pp 159-182.

6. Kester, J.J.; Fennema, O.R.; *Food Technol.* **1986**. *48*, 47-59.

7. Kamper, S. L.; Fennema, O.;. *J. Food Sci.* **1985**, *50*, 382-384.

8. Kester, J.J; Fennema, O.R.;. *J. Food. Sci.*, **1989**, *54*, 1383-1389.

9. Nussinovitch, A.; Hershko, V.;. *Carbohydrates polymers*. **1997**, *30*, 185-192.

10. Sensidoni, A.; Peressini, D.; Callegarin, F.; Debeaufort, F.; Voilley, A.; *J. Food Sci.* **1998**, In press.

11. Balasubramaniam, V.; Chinnan, M.S.; Mallikarjunan; P.; Phillips; R.D. *J. Food Process Eng.* **1997**, *20*, 17-29.

12. Peyron, A ; *Viande. Produits carnés.* **1991** , *12*, 123-127.

Chapter 3

Utilization of Antimicrobial Packaging Films for Inhibition of Selected Microorganisms

K. Cooksey

Department of Packaging Science, Clemson University,
228 Poole Agricultural Center, Clemson, SC 29634

Antimicrobial additives can be successfully incorporated into food packaging films or film coatings. However, several factors affect their effectiveness for inhibition of spoilage or pathogenic microorganisms. Nisin was coated onto low density polyethylene film using a cellulose based coating. It inhibited growth of *Staphylococcus aureus* and *Listeria monocytogenes* within 48h in a nonfood system. Other studies using different coating material, methods and antimicrobial agents are reviewed. Several factors affecting the effectiveness of antimicrobial packaging materials are also discussed.

Surface growth of microorganisms is one of the leading causes of food spoilage. Natural microflora can eventually spoil the food or surfaces can be contaminated by handling during processing and packaging. For many years foods have been treated with antimicrobial agents however, packaging materials may also provide the same benefits using similar or different additives. A packaging system that allows for slow release of an antimicrobial agent into the food could significantly increase the shelf life and improve the quality of a variety of foods. The use of these packaging systems are not meant as a "cover up" for poor quality control. It can, however, serve as an additional protective measure to help ensure safe and high quality foods.

The purpose of this paper is to provide information regarding the types of antimicrobial films used, their effectiveness and factors that influence their effectiveness. First a specific study done in our laboratory using nisin as an antimicrobial agent coated onto LDPE film will be covered. Then a review of other studies using different types of antimicrobial packaging systems follows. Finally the factors that affect the implementation of an antimicrobial packaging system will be discussed.

17

Effectiveness of Nisin Coated LDPE Film

A study was done to find a way to coat LDPE film with a material containing nisin and test the effectiveness of the film for inhibition of *Staphylococcus aureus* and *Listeria monocytogenes*. *(1)*. Nisin is a bacteriocin produced by *Lactococcus lactis* and is considered a natural additive, which is considered to be a desirable feature for many food additives. It has GRAS status for application with processed cheese and is particularly effective for preventing growth of *Clostridium botulinum*. Nisin is selective for gram positive bacteria but isn't effective for inhibition of gram negative bacteria. *S. aureus and L. monocytogenes* are both gram positive pathogenic microorganisms.

Methodology The film coating solution was made using 7.0g methylcellulose (SIGMA, St. Louis, MO), 3.0g hydroxypropyl methylcellulose (SIGMA, St. Louis, MO), 100mL distilled water, and 200mL 95% ethanol. Methylcellulose (MC) and hydroxypropyl methylcellulose (HPMC) were slowly added to distilled water and stirred until completely dissolved. Ethanol was slowly added to the solution and 6mL of polyethylene glycol (PEG) (SIGMA, St. Louis, MO), was added as a plasticizer.

Nisin (SIGMA, St. Louis, MO), was prepared using 2.0g in 10.0mL of 0.02N hydrochloric acid making it a 200IU/mL concentration. Additional solutions were made using 1.5 g (150 IU/mL), 1.0 g (100 IU/mL), 0.5g (50 IU/mL) in 10mL of 0.02N hydrochloric acid.

Four different coating solutions were made containing 200, 150, 100, 50 and 0 IU/mL nisin. The coating solution was applied to low density polyethylene film (1.5 mil) using a thin layer chromatography plate coater set at a thickness of 500μm. Glass plates (8 x 8 in) were covered with LDPE film and passed through the TLC plate coater. Each coated material was dried overnight at room temperature. The film samples were cut into 2 inch (25cm^2) squares and sterilized by UV light for 2 min prior to inoculation with *S. aureus* or *L. monocytogenes*. Since films had been handled during formation and cutting, some contamination could have occurred. Therefore, a preliminary test was done 1, 2 and 3 mins of UV light exposure. Two minutes was found to be sufficient and was used in the study.

After film sterilization, the samples were soaked separately with solutions containing 10^6 CFU/mL *S. aureus* and 10^5 CFU/mL *L. monocytogenes*. All film samples were stored at room temperature for 24 h to allow nisin to take action on the bacteria. After 24 h the film (coated side down) was transferred onto Tryptic Soy Agar plates and incubated at 35°C for 48h.

Colonies were selected for identification. *S. aureus* was confirmed using a gram stain and *L. monocytogenes* was confirmed by growth on modified oxford agar and formation of black colonies. The data was reported as CFU/25 cm^2 based on the area of the film sample. The tests were performed in triplicate.

Results   The results of the film samples inoculated with *S. aureus* and *L. monocytogenes* are shown in Table I. Low levels of inoculum were transferred onto the film, probably because of the low moisture content on the film surfaces. Nisin levels of 100 IU/mL were effective for inhibiting *S. aureus* on the surface of the coated LDPE film samples. Even lower levels of *L. monocytogenes* inoculum were transferred onto film compared to the samples inoculated with *S. aureus*. The population of *L. monocytogenes* tested with film samples containing 50-150 IU/mL nisin varied between the three replications, therefore no trend could be determined. However, it is clear that 200 IU/mL was effective for inhibiting the growth of *L. monocytogenes*.

Table 1. Inhibition of *S. aureus* and *L. monocytogenes* using nisin coated
LDPE film

Levels of nisin (IU/mL)	Levels of *S. aureus* (CFU/ 25cm^2)	Levels of *L. monocytogenes* (CFU/ 25cm^2)
0	160	21
50	0.3	2
100	0	18
150	0	5
200	0	0

Conclusions of nisin coated LDPE film study   LDPE film was successfully coated with nisin using MC/HPMC as the carrier. Low levels of inoculum observed on the film samples was probably due to low moisture content of the film. Nisin was very effective for inhibition of *S. aureus* but required a higher concentration for inhibition *of L. monocytogenes*. Although no specific tests were done, there was a noticeable decrease in the clarity of the film and heat sealing with impulse and hot bar sealers was difficult. Therefore more work will be done to establish the physical characteristics of film coated with MC/HPMC containing nisin.

Other Studies Regarding Antimicrobial Food Packaging Systems

The study of antimicrobial food packaging films has shown that additives can be very effective for reducing surface microbial growth and possibly extend the shelf life of shelf stable and fresh refrigerated foods. However there are many factors to consider when selecting an antimicrobial packaging system. The type of antimicrobial additive, how to control the release of the additive and factors that influence the effectiveness of the additive must be examined.

In general, the antimicrobial systems can be broken down into two categories. The first is the direct incorporation of the additive into a packaging material, either within the matrix or as a coating. This method depends on intimate surface contact with the food for effectiveness. The second category is an indirect method, and doesn't rely on direct contact. For example, insertion of a component (such as an oxygen sachet) within the package which interacts with the food to reduce spoilage. In both cases the end result is extended shelf life and both can be very effective.

There are many different types of antimicrobial additives. The most popular are the naturally derived additives due to their perceived consumer "friendly" standing. The types of additives and their modes of action vary widely. Some antimicrobial agents include benzoic acid, sodium benzoate, sorbic acid, potassium sorbate and propionic acid. Many of these substances have been used commercially during food processing but they have also been used with edible coatings (2). Other agents have included bacteriocins and plant derived compounds. The following is a review of studies using different types of antimicrobial additives and the factors that influence their effectiveness.

Direct incorporation of antimicrobial additives A very important factor in the direct incorporation of an antimicrobial additive into or onto a packaging film, is establishing which matrix to use as the "carrier" for the additive. This can be one of the most difficult tasks. It is important to contain the additive yet allow it to disperse onto the food surface.

Vojdani and Torres (3) chose polysaccharide films such as chitosan, methylcellulose (MC) and hydroxypropyl methylcellulose (HPMC) as a carrier for potassium sorbate. They indicated that sorbate eventually absorbs into the food from the surface and therefore, the protective effect is lost. The film coatings could help retain the levels of sorbate at the surface of the food depending on how much potassium sorbate diffused through the film. They determined that all polysaccharide films with sorbate could improve surface protection of foods with an estimated effective durability of 1.5-2.0 months at room temperature and 3.5-5 months at refrigeration temperature. Chitosan was found to have the highest permeation while the methylcellulose and hydroxypropyl methylcellulose combination had the lowest. Films with the lowest permeation were considered the most desirable because it allows sorbate to diffuse to the surface of the food over a longer period of time.

Reduction in the permeation of potassium sorbate with the use of fatty acids was achieved in further studies by Vojdani and Torres (4). A single layer of MC/HPMC coated with another layer of MC/HPMC containing different fatty acids such as lauric, palmitic, stearic and arachidic was tested. MC/HPMC film was used as a base layer for a coating of hot solution containing lipids or edible wax. Embedded films were made by sandwiching the lipid layer between two layers of MC/HPMC film. Their findings indicated that MC/HPMC films coated with a second layer of MC/HPMC containing beeswax had the lowest permeation of potassium sorbate and therefore could be used to increase the shelf life of refrigerated or shelf stable foods. Of the films using fatty acids, palmitic acid provided the lowest permeation *(5)*.

Wong et al., *(6)* also recognized the importance of creating a matrix for control of sorbate release onto a food surface. The matrix was created using a calcium set alginate gel using a pH controlled room temperature formation method. They also made the same film except the calcium alginate film was prepared hot. In general, the pH controlled prepared film had a lower permeability rate (1.06×10^{-7} cm^2/sec) than the hot prepared film (1.58×10^{-7} cm^2/sec). They theorized this effect was due to an ordered, controlled network formation created by the pH controlled method of producing calcium alginate film. The hot method created a homogenous film but was rougher and more prone to leakage. A more ordered network would be advantageous for the controlled release of a preservative over time.

The studies discussed so far used edible films as the matrix for retention and release of antimicrobial additives but others have worked with synthetic polymers as a matrix. Weng and Hotchkiss *(7)* established a method for incorporation of an antifungal agent called imazalil for reduction of surface spoilage in cheese. Incorporation of benzoic anhydride (0.5%) into LDPE was successful and reduced the growth of *Aspergillis toxicarius* and *Penicillium* spp. by conversion into benzoic acid. *Rhizopus stolonifer* growth was completely inhibited by 0.5% benzoic anhydride LDPE film while 0.1% was needed for complete inhibition of *A. toxicarius* and *Penicillium* spp *(8)*.

A plant derived antimicrobial additive called allyl isothiocyante (AIT) has been incorporated in food packaging materials. AIT is an approved additive in Japan and provides antibacterial action by diffusing from the packaging material as a vapor which can surround the food *(9)*. It has been effective for increasing the shelf life of meat, fish and cheese at levels of 34-110 ng/mL for bacteria and 16-62 ng/mL of yeast and mold *(10)*. Lim and Tung *(11)* studied the permeation of AIT vapor through polyvinylidene chloride/polyvinyl chloride copolymer films. At a fixed vapor pressure, AIT exhibited increased permeation and diffusion with increasing temperature but solubility decreased. The authors provided a method to predict permeation which could be used in future studies to design the best packaging material for permeation of AIT or other permeants.

Another method of incorporating an antimicrobial additive to synthetic food packaging materials involves spraying the surfaces of the package with a powder containing an antimicrobial agent. Ming et al., *(12)* produced a bacteriocin-rich powder which was used to inhibit the growth of *Listeria monocytogenes* in solution and in meats. Nisin and pediocin activity was retained in a dried milk powder base but pediocin was considered to retain more of its activity. Pediocin/milk powder base was applied to the inside of vacuum barrier bags containing fresh turkey, processed ham or fresh beef inoculated with *L. monocytogenes*. Growth of *L. monocytogenes* was prevented in packages with ham while population levels decreased steadily over the 12 week refrigerated storage period for fresh turkey and beef packages. Their study proved that a package dusted with an antimicrobial agent can be very effective for inhibiting the growth of pathogenic bacteria in meats.

Other systems that could be incorporated into a packaging material include enzymes that could release antimicrobial by products such as hydrogen peroxide which can provide a preservative effect *(13)*. Carbon dioxide can also be incorporated into a polymer film matrix and slowly released. Other additives include zeolite (release of silver ions), benomyl and hinokitiol also known as and ß-thujapricin which is derived from cypress bark *(14)*.

Indirect incorporation of antimicrobial additives The use of sachets has been an indirect method of controlling the microbial growth on food surfaces in packages. Oxygen absorbing materials encased in a permeable material have been used commercially for many years. Films may also be made with oxygen absorbing properties within the polymer matrix, thus eliminating the need for the packet commonly inserted in packages today.

A packet containing silica with entrapped ethanol can also be used to release an ethanol vapor within a package *(13)*. A Tyvek pouch containing sorbitol was inserted in a PVC overwrapped tray containing mushrooms *(15)*. Packages containing 15g sorbitol had significantly lower total plate counts than packages without sorbitol stored for 6 days at 12°C.

Factors to Consider for Commercial Implementation

The type of carrier used and antimicrobial agent used are important in developing an antimicrobial packaging system. However when used with foods, a variety of factors can alter the effectiveness of the system to protect the food.

Water activity Vojdani and Torres *(3)* found that measuring the permeation of potassium sorbate through polysaccharide films was significantly affected by the water activity of the film tested. A higher A_w increased permeation which has a negative effect on the amount of potassium sorbate available for surface protection. This was further confirmed using MC/HPMC film containing palmitic acid that had much higher permeation rates at A_w 0.75-0.80 compared to 0.70 and 0.65. *(16)*.

These results were confirmed by the findings of Wong et al., *(6)*. They also observed a trend of increasing permeability with increasing A_w. The exact reason for this observation was not clear. Wong et al., *(6)* indicated that increased film hydration would also increase the chances for solution to come into contact with the film. However, they also said that at a lower A_w the film is less hydrated and more compact. Therefore the distance for the solute to travel through the film would be less and the permeation would increase with decreasing A_w. However, this effect was not observed in their study.

pH According to Rico-Pena and Torres *(16)*, the permeation of sorbic acid was affected by the pH of the of it's environment. Permeation of sorbic acid decreased as pH increased from 3.0 to 7.0 when measured at an A_w of 0.80. Weng and Hotchkiss *(8)* also found benzoic anhydride LDPE film was more effective for inhibition of molds at a lower pH.

Temperature Several researchers have found that the protective effects of the antimicrobial films become less effective at higher temperatures. This is not too surprising since the diffusion of the antimicrobial agents from it's matrix would be expected to increase at a higher temperature. *(3,4,6)*. A careful balance of diffusion must be maintained so that enough of the agent can diffuse to be effective but not so much that it no longer remains on the surface of the food and becomes ineffective. Temperature is one of the most important factors to control. Weng and Hotchkiss *(8)* stated that lower levels of benzoic anhydrides in LDPE might be as effective at refrigeration temperature compared to the higher levels used at room temperature.

Chemical interaction with film matrix Different additives will permeate through a polymer matrix at different rates depending on the nature of their interaction within the matrix. This can include molecular weight, ionic charge and solubility. An example of varying permeability based on ionization state is provided by Wong et al., *(6)*. They compared the permeation of ascorbic acid, potassium sorbate and sodium ascorbate in calcium-alginate films at 8, 15 and 23°C. They found that ascorbic acid had the highest permeation compared to potassium sorbate. Sodium ascorbate had the lowest permeation. The same effect was observed at all temperature ranges. The authors felt the ionic state of the different additives was the reason for the effect observed.

One important factor for incorporation of additives with low density polyethylene (LDPE) is the polarity and molecular weight of the additive. Since LDPE is nonpolar, use of additives with a higher molecular weight and lower polarity would be more compatible with this material *(8)*. Organic acids such as sorbic acid, propionic acid and benzoic acid did not incorporate into LDPE or did not migrate in sufficient amounts to prevent to growth of mold. However, incorporation of benzoic acid as an anhydride was found to be effective and therefore was more compatible for use with LDPE film *(8)*.

Physical properties of packaging material Little research has been done on the physical properties of films with antimicrobial additives either incorporated into the film matrix or as a coating. Some studies have been done on properties of edible films that could provide insight into the characteristics of antimicrobial films made with similar components. It would be reasonable to assume that the flexibility, tensile and elongation of a film could be affected as evidenced by research with edible films. Weng and Hotchkiss (8) reported no noticeable differences in clarity or strength of LDPE film containing benzoic anhydride. However, LDPE film coated with MC/HPMC containing nisin became difficult to heat seal after coating with MC/HPMC containing nisin (1).

Cost Specific costs related to films coated with antimicrobial agents could not be found however, the cost of edible films would be a reasonable comparison. According to Fishman (17) cellulose based films cost from $4.50-7.00/lb. and pectin-containing films cost from $5.00-8.00. Supplementation with starch and glycerol can reduce the cost of pectin-containing films by as much as 50%. (17). When compared to the cost of basic polymer films these costs are significantly high enough to make the commercialization of such films attractive only for high value food products.

FDA Approval Organic acids and some bacteriocins have FDA approval as additives for some foods. However, plant extracts, such as AIT, are not currently approved in the U.S. The reason why AIT is not approved is because of the safety concern regarding a toxic material that can be produced during the manufacturing of synthetic AIT. As mentioned before, AIT is approved in Japan.

Weng and Hotchkiss (8) also indicated that benzoic anhydride was not a FDA approved additive at the time it was used in their study. On the other hand, benzoic anhydride was converted into benzoic acid, which is an approved additive. The approval of silver ions (zeolite) as an additive in the U.S. is also under review (18).

Conclusion

Much of the work reviewed provides evidence that antimicrobial agents are effective for inhibition of microbial growth when incorporated into and/or onto packaging materials. More work is needed with regard to the factors that affect the inhibitory action of the antimicrobial systems. Little has been done with regard to the sensory effects these additives have on the food, particularly with regard to plant extract additives, which can give off pungent odors. The effects of the additives on the physical characteristics of the film also needs further study. One of the keys to commercialization of antimicrobial packaging systems is cost effective production. Partnerships between researchers and packaging converters can help achieve this goal. Finally, we must remember that the primary purpose of an antimicrobial packaging system is to help provide safe and healthy food and is to be used along with high quality control standards.

Literature Cited

1. Cooksey, K. *J. Food Sci.* **1999**, (manuscript in preparation).
2. Cuppett, S.L. In *Edible Coatings and Films to Improve Food Quality*, Krochta, J.M.; Baldwin E.A. and M. Nisperos-Carriedo, Eds., Technomic Pub. Co. Lancaster, PA, **1994**, pp.121-137.
3. Vojdani, F. and Torres, J.A. *J. Food Process Eng.* **1989**, 12, pp.33-48.
4. Vojdani, F. and Torres, J.A. *J. Food Process Eng.* **1989**, 13, pp.417-430.
5. Vojdani, F. and Torres, J.A. *J. Food Sci.* **1990**, 55, pp.841-846.
6. Wong, D.W.S.; Gregorski, K.S.; Hudson, J.S. and Pavlath, A.E. *J. Food Sci.* **1996**, 61, pp.337-341.
7. Weng, Y.-M. and Hotchkiss, J.H. *J. Fd. Protect.* **1992**, 55, pp.367-369.
8. Weng, Y.-M. and Hotchkiss, J.H. *Packaging Tech. and Sci.* **1993**, 6, pp.123-128.
9. Delaquis, P.J. and Mazza, G. *Food Tech.* **1995**, 49, pp.73-84.
10. Isshiki, K., Tokuora, K., Mori, R. and Chiba, S. *Biosci. Biotech. Biochem.* **1992**, 56, pp.1476-1477.
11. Lim, L.T. and Tung, M.A. *J. Food Sci.* **1997**, 62, pp.1061-1066.
12. Ming, X., Weber, G.H., Ayres, J.W. and Sandine, W.E. *J. Food Sci.* **1997**, 62, pp.413-415.
13. Labuza, T. *Food Tech.* **1996**, 50, pp.68-71.
14. Rooney, M.L. In *Active Food Packaging* , Rooney, M.L., Ed.; Blackie Academic and Professional, London, **1995**, pp.74-110.
15. Roy, R., Anatheswaran, R.C. and Beelman, R.B. *J. Food Sci.* **1995**, 60, pp.1254-1259.
16. Rico-Pena, D.C. and Torres, J.A. *J. Food Sci.* **1991**, 56, pp.497-499.
17. Fishman, M. *Food Tech.* **1997**, 51, p.16.
18. Hotchkiss, J.H. In *Active Food Packaging* , Rooney, M.L., Ed.; (1995). Blackie Academic and Professional, London, **1995**, pp.238-255.

Chapter 4

Modeling of Additive Diffusion Coefficients in Polyolefins

J. Brandsch, P. Mercea, and O. Piringer

FABES Research Inc. for Analysis and Evaluation of Mass Transfer,
Schragenhofstrasse 35, D–80992 Munich, Germany

An equation for predicting the diffusion coefficients of hydrocarbons in polyolefins is developed. The calculated values are compared with experimental results obtained with n-paraffins and organic compounds of different polarities and structures. The possibilities and limitations of the model are discussed in the light of the regulations for additives in food contact materials.

One of the principal aims of the regulations for food contact materials and articles is the protection of the consumer. Of great help in the evaluation procedures is the prediction of migration based on the theory of diffusion. All that is needed for a reasonable prediction of migration in many practical cases is the availability of data for two fundamental constants: the partition coefficient $K_{P/L}$ of a migrating solute between the plastic P and the foodstuff or simulating liquid L and the diffusion coefficient D_P of the solute in P. Assuming that data for $K_{P/L}$ and D_P do exist or can be predicted with sufficient accuracy, a considerable reduction in analytical work, time and financial support would be possible.

The following considerations refer to the estimation of values for D_P and their use for prediction of migration from plastics. Special consideration is given to the comparison of predicted values with experimental results obtained with n-paraffins from polyolefins.

For practical purposes most estimation methods for D_P are quite complex, require numerous parameters themselves and until now have been suitable only for estimates of diffusion coefficients of gases in polymers. The development of estimation models for diffusion is hindered by a lack of fundamental understanding of the interrelationships of density, orientation and morphology of the polymer and the chemical nature, size, shape and concentration of the diffusing molecule. Furthermore, the diffusion coefficient can also be affected by the penetration of foodstuffs or simulating liquids into the polymer.

One solution for the estimation of a diffusion coefficient is to make correlations of the diffusion coefficient with the relative molecular mass M_r of the solute for each polymer system and temperature T of interest. This approach has already been successfully used (1,2). With an empirically obtained coefficient A_P which accounts for the effect of the polymer on diffusivity, a simple equation for the estimation of D_P as a function of A_P, M_r and T has been proposed (2-4). The aim of the following considerations is to generate a theoretical base for this equation and to provide a comparison of predicted D_P-values with the best experimental data that are available from the literature. It should be emphasized that the experimental measurement of D_P-values for solutes with higher molecular masses is difficult which, as a consequence, explains that some experimental data found in the literature are questionable.

Modeling of properties in homologous series

In many areas of research a very useful means for prediction of properties is to start with homologous series of chemical compounds where the structures of the individual members of the series do not deviate significantly from one another. Out of all known classes of chemical compounds the saturated open-chain and unbranched (normal) hydrocarbons, the n-paraffins, represent the most frequent studied homologous series with the largest number of available members in pure form. The number i of carbon atoms in a n-paraffin molecule, or more exactly, the number of methylene groups including the two methyl groups can be interpreted as playing the role of i identical interacting structure subunits that compose the molecule. A property of a macroscopic n-paraffin sample can then be described as a function of the i subunits which are essentially the identical subunits making up every molecule of the macroscopic system. These characteristics of a homologous series provide the theoretical basis for a behavior which can be described by an asymptotic correlation. Many such correlations are known in the literature. A general equation for correlating thermophysical properties of n-paraffins was proposed recently (5). With a number of adjustable parameters which are evaluated by fitting data for lower carbon numbers, the values of a property for the whole homologous series can be calculated.

Another approach uses an i-fold multiplicative combination of a magnitude, leading to an asymptotic limit value for i>>1. Following this way a function $f(i) = c_0$ $w_{i,e} = c_0(1+2\pi/i)^{i/e}$ has been derived (6) with a limit value $f(i \to \infty) = c_0 e^{2\pi/e} = c_0 w$ and w = 10.089. By c_0 we designate a specific constant, for example with the dimension of a molar energy, that is valid for all terms with the analogous structure of a homologous sequence. It is therefore possible to assign any term of the power sequence $w_{i,e}$ to the corresponding value $f(i)$ of a property of the member i in the homologous sequence of chemical compounds. The ratio between $f(i)$ and the asymptotically reached limit $f(\infty)$ for i>>1 is dimensionless and independent of c_0:

$$\frac{f(i)}{f(\infty)} = \frac{c_0(1+2\pi/i)^{i/e}}{c_0 e^{2\pi/e}} = \frac{w_{i,e}}{w} \tag{1}$$

As a consequence the value of c_0 must not be known so far it can be assumed to be the same for different members of the homologous series. By e in the exponent in equation 1 and in the subscript of $w_{i,e}$ the transcendental number 2.71... is designated.

In the following an application of equation 1 is given to be understood as an important step in the modeling of diffusion coefficients.

The critical temperatures of n-paraffins

The critical temperature may be considered to be a measure of the intensity of interaction between the n molecules of a system. This system is free of significant dipole and hydrogen bonding interactions and the intensity of interaction is produced by „van der Waals" forces. Although the critical temperature of a macroscopic system is practically independent of the number of molecules, there is a possibility to estimate the influence of the number i of structural subunits out of which a molecule is made up, on the value of the critical temperature of a macroscopic system. The critical temperatures are especially suitable for comparison of the numerical values within a homologous sequence because at these temperatures the systems are in corresponding states.

If we designate by $T_{c,i}$ and $T_{c,\infty}$ the critical temperatures of paraffins containing a number i and j→∞ of carbon atoms, respectively, an asymptotic correlation between the critical temperatures of the homologous paraffins and the power sequence $w_{i,e}$ can be established using equation 1 (6):

$$\frac{T_{c,i}}{T_{c,\infty}} = \frac{(1 + 2\pi / i)^{i/e}}{e^{2\pi/e}} \qquad (2)$$

Experimental values of the critical temperatures of n-alkanes are known up to eicosane (i=20)(7).

Table I: Critical Temperatures of the n-Paraffins with i Carbon Atoms

Number i of carbon atoms	$T_{c,i}$ / K measured	$T_{c,\infty}$ / K calculated	$T_c - T_{c,\infty}$ [a]
9	594.6	1039.1	- 2.9
10	617.7	1036.8	- 0.6
11	638.8	1035.5	+ 0.7
12	658.2	1034.9	+ 1.3
13	676.0	1034.9	+ 1.4
14	693.0	1036.0	+ 0.2
15	707	1034.7	+ 1.5
16	722	1036.7	- 0.5
17	733	1034.4	+ 1.8
18	748	1039.2	- 3.0
19	756	1035.4	+ 0.8
20	767	1036.8	- 0.6

[a] $T_c = 1036.2$ K represents the mean value of the 12 calculated $T_{c,\infty}$ - values

By means of equation 2 it is possible to calculate, starting from each experimental value corresponding to i carbon atoms, a limit value $T_{c,\infty}$ (Table I).

Due to the fact that in the initial members of the n-paraffin sequence the terminal methyl groups contribute to a more important deviation from the system which is basically made up only of methylene groups, it is more convenient for the determination of $T_{c,\infty}$ to use paraffins with long chains. As may be seen from Table I, these deviations become insignificant for $i \geq 9$ because the individual values scatter around the mean value $T_c = 1036.2$ K, obtained from $T_{c,\infty}$ values corresponding to the 12 longest chains (i=9 to 20). Figure 1 shows the trend of the curve resulting from equation 2 with $T_{c,\infty} = T_c = 1036.2$ K in comparison with the experimental values $T_{c,i}$ for $1 \leq i \leq 20$.

The remarkable coincidence between the experimental values of the critical temperatures within the homologous sequence and the predicted values with equation 2, using only one adjustable parameter, the limit value T_c, sustains the interpretation of $w_{i,e}$ as a measure of the relative density of the interaction energy in a condensed macroscopic system of identical molecules in form of chains with a number of i identical substructure units.

An equation for the diffusion coefficients of n-paraffins in polymethylene

The molar volume V_m of a perfect gas as well as its coefficient of selfdiffusion D_G are inversely proportional related to the pressure p. For a gas at constant temperature the following relationship is therefore valid:

$$V_{m,2}/V_{m,1} = p_1/p_2 = D_{G,2}/D_{G,1} = \exp(\Delta S/R) = e^c \qquad (3)$$

where ΔS, R and c stands for the difference in the molar entropy of the states 1 and 2, the gas constant and a dimensionless number, respectively.

Let us now consider a macroscopic system in an amorphous state above its glass temperature. The particles of the system are n-paraffins with a number $i \gg 1$ of carbon atoms in the molecular chain. Let us first consider the theoretical case with one single macromolecule of infinite length as a polymethylene chain in form of a disordered coil. Due to the possibility of free rotation of any of the methylene subunits around the bond axis to the neighbour subunits, a relative motion of segments of more subunits is also possible.

In the following an equation for the diffusion coefficient $D_{P,i}$ of a n-paraffin with i carbon atoms in a polymethylene chain of infinite length can be derived within 3 approximation steps:

Neglecting in the first approximation both, the existence of an activation energy E_A of the diffusion process and the influence of the volume and mass of the diffusing solute, a ratio of diffusion coefficients D_2/D_1 is postulated as an exponential function in conformity with equation 3 using the value $w = e^{2\pi/c}$ for c in equation 3: $D_2/D_1 = \exp(w) = \exp(wR/R) = \exp(\Delta S_w/R)$. Related to a value D_1 for an initial state 1 the amount of D_2 is a measure of the disordered motion of the methylene groups, with a corresponding increase of the molar entropy $\Delta S_w = Rw$, resulting from the interaction

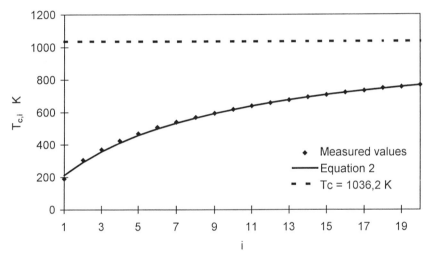

Figure 1. Critical Temperatures of n-Paraffins as a Function of the Number i of Carbon Atoms.

between these groups in the polymer matrix with the relative density of interaction energy w. One mole of polymethylene is defined as one mole of methylene groups, - CH_2-. The disordered motion of the methylene groups providing the value D_2 related to D_1 is assumed as analogous to the reversible expansion in a perfect gas with the same change in entropy ΔS_w.

In a second approximation a molar activation energy E_A of the motion of the methylene groups in the polymethylene chain is assumed. This activation energy $E_A = wRT_c = 10.089 \times 8.31451 \times 1036.2 = 86.923$ kJ mol^{-1} is defined as a magnitude proportional to w and the limit value of the critical temperature $T_c = 1036.2$ K in the homologous series of n-paraffins. In this way $D_2/D_1 = e^w \exp(- E_A/RT)$. With this expression we have $D_2 = D_1$ for $T = T_c$ and this amount is defined as the unit value $D_0 = 1$ $m^2 s^{-1}$.

The third approximation takes into account the diffusing solute as a n-paraffin with a number i of carbon atoms in its molecular chain. The molar volume of a compound designated by b as in the van der Waals equation, may be calculated by means of the critical molar volume $V_{m,c}$, because $b=V_{m,c}/3$. Let be $A_{m,i} = b_i^{2/3}$ $m^2 mol^{-1}$ the molar cross-sectional area of a diffusing solute. By moving through the polymer matrix this cross section must overcome the force exerted on it by the methylene groups in the polymer matrix during their disordered motion. The value of this force F divided by the unit area $A_0 = 1$ m^2 to which the force is applied defines a pressure p = F/A_0. The product $pA_{m,i}d = (A_{m,i}/A_0)Fd = E_{m,A,i}$ J mol^{-1} represents the necessary molar work to overcome the resistance of the matrix by moving $A_{m,i}$ along the distance d in a direction perpendicular to it. Referred to the corresponding work for moving A_0 along the same distance, $E_{0,A} = Fd$ J, we get $E_{m,A,i} /E_{0,A} = \varepsilon_{m,A,i} = A_{m,i} /A_0 = A_{m,i}$ mol^{-1}. On the other side, the pressure p can be expressed as $p = m(N/V)<v^2>$, with the density N/V of the moving matrix subunits (methylene groups), the subunits mass $m = M_r m_u$ and the mean value of the square of the subunits velocity $<v^2>$, respectively. With the atomic mass unit m_u, the relative subunits mass M_r, the amount of subunits n (mol), the Avogadro constant N_A and the volume V, we get $p = M_r m_u(nN_A/V)<v^2> = m_u N_A<v^2>M_r/(V/n)$. The product $p(V/n) = m_u N_A M_r<v^2> = 10^{-3}M_r<v^2> = E_{m,k}$ J mol^{-1} defines a molar kinetic energy of the moving subunits. Referred to the corresponding energy $E_{0,k} = m_0 M_r<v^2>$ J, using the mass unit $m_0 = 1$ kg instead of the molar mass unit $M_u = m_u N_A = 10^{-3}$ kg mol^{-1}, we get $E_{m,k}/E_{0,k} = \varepsilon_{m,k} = 10^{-3}$ mol^{-1}.

From the above considerations a dimensionless value $\varepsilon_{m,i} = \varepsilon_{m,A,i}/\varepsilon_{m,k} = 1000 A_{m,i} = 1000$ $b_i^{2/3}$ results as a relative measure of the resistance against the movement of the diffusing solute. This value is used as a further negative term in addition to the activation energy. As a final result the following equation can be established for the diffusion coefficient $D_{P,i}$ of a n-paraffin with i carbon atoms in an amorphous matrix of polymethylene:

$$D_{P,i} = D_0 \exp\left[w - 1000 (V_{m,c,i}/3)^{2/3} - E_A / RT\right] =$$
$$= 10^4 \exp(w - 0.13 M_{r,i}^{2/3} - \frac{10450}{T})\ \ cm^2/s \tag{4}$$

Due to the repeating -CH$_2$- structure in the n-paraffins, a constant value of the ratio $V_{m,c,i}/M_{r,i}$ can be expected for this homologous series excepting the first few members with a substantial „impurity" of this substructure. For the n-paraffins with i = 5-17 a mean value $V_{m,c,i}/M_{r,i}$ = 4.24E-6 m^{3}mol^{-1} is obtained (7). The constant 0.13 in equation 4 results from $1000(4.24 \times 10^{-6}/3)^{2/3} \approx 0.13$.

Equation 4 can be used as a reference equation for all polyolefins. It represents a theoretical construct resulting from an asymptotic correlation and an assumed infinite chain of methylene groups representing the amorphous polymer matrix.

Whereas w = $e^{2\pi/e}$ = 10.089 = A_P stands for the theoretical structure of polymethylene, other characteristic A_P-values can result for technical products depending of their specific structure in the polymer matrix. Nevertheless the remaining two terms in the exponent of equation 4 can be hold unchanged for polyolefins and paraffins. For other diffusing compounds the corresponding critical molar volumes could be more appropriate as the molecular weights. Understanding A_P as a characteristic structural parameter of the polymer which must be determined experimentally, the following more general equation for the diffusion coefficient $D_{P,i}$ can be used:

$$D_{P,i} = 10^4 \exp(A_P - 0.13\, M_{r,i}^{2/3} - \frac{10450}{T}) \qquad cm^2/s \qquad (5)$$

The factor 0.13 in equation 5 can be used as an acceptable approximation for most hydrocarbons and other nonpolar solutes.

Comparison of calculated and experimental data

The diffusion coefficients of n-paraffins with 12 to 22 carbon atoms in high density (HDPE) and low density polyethylene (LDPE) have been measured by a permeation method (8). Methanol (MeOH) and ethanol (EtOH) were used in order to avoid interaction between these polar solvents and the nonpolar polymers. Indeed with both polar solvents no interaction occurred in the investigated temperature range between 6 and 40 °C.

Figure 2 contains the measured values of the diffusion coefficients from HDPE and LDPE at 23 °C and the calculated curves obtained with equations 4 and 5 in the corresponding range of masses. The measured diffusion coefficients are in good agreement with the calculated values obtained with equation 5 using A_P = 8.8 for HDPE and A_P = 10.6 for LDPE, respectively. The most important finding resulting from this representation is the agreement of the experimental values with the $M_{r,i}^{2/3}$-dependence in the exponent of equation 5 and the reference equation 4 with the theoretical value A_P = w = 10.089.

Figure 3 shows the temperature dependence of the diffusion coefficients obtained with i = 12 to i = 22 in LDPE. The comparison of the experimental data with the corresponding curves obtained with equation 5 and A_P = 10.6 for i = 12 and i = 22 shows again a reasonable agreement. This result is used as a proof for the order of

34

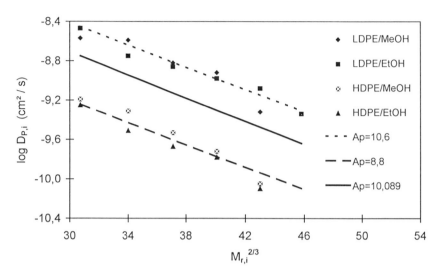

Figure 2. Logarithm of Diffusion Coefficients of n-Paraffins in Polyolefins at 23 °C as a Function of the Relative Molecular Mass.

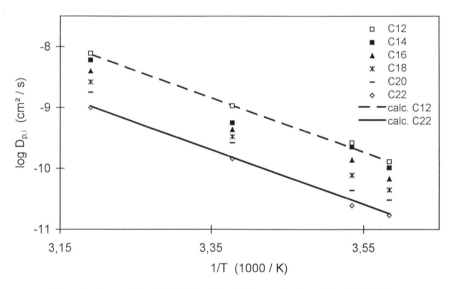

Figure 3. Logarithm of Diffusion Coefficients of n-Paraffins in LDPE as a Function of Temperature.

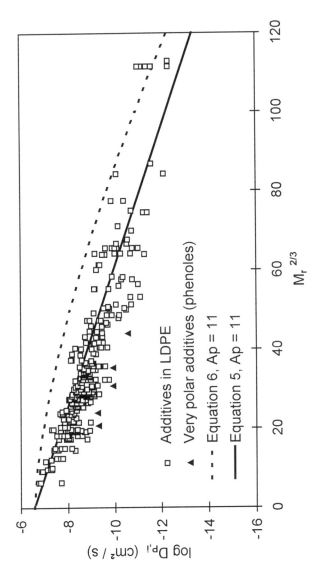

Figure 4. Logarithm of Diffusion Coefficients of a Variety of Additives in LDPE at 23 °C as a Function of the Relative Molecular Mass.

magnitude for the activation energy E_A used in the reference equation 4 and defined before as $E_A = wRT_c = 86.923$ kJ mol^{-1}.

From a data collection of more as 300 measured diffusion coefficients in LDPE and LLDPE (linear LDPE) at 23 °C, containing solutes with different molecular masses and molecular structures, a distribution as shown in Figure 4 has been obtained. The data cluster around the curve obtained with equation 5 and $A_P = 11$. It must be emphasized that experimental values for one solute obtained in different laboratories, measured with different methods and using different sources for the same polymer type, scatter often more than one order of magnitude. A source of big errors is a strong interaction of solvents with the polyolefins, especially at high temperatures. A structure depending deviation of measured diffusion coefficients towards lower values in comparison with the calculated values using equation 5 results especially with high polar solutes, for example with phenols, as indicated in Figure 4.

Consequences for Food Regulation. For the protection of the consumer the most important result from theoretical predictions is the possibility to indicate an upper limit for the $D_{P,i}$-values. One can see from Figure 4 that the previously (*2-4*) proposed equation

$$D_{P,i} = 10^4 \exp(A_P - 0.01\, M_{r,i} - 10450/T) \quad \text{cm}^2/\text{s} \tag{6}$$

provides such an upper limit. It is only a matter of convention to select an appropiate A_P-value in order to avoid underestimation of $D_{P,i}$-values in less as 1 or 5 % of the cases which occur in practice. In Figure 4 the curve obtained with equation 6 using $A_P = 11$ for LDPE shows that all experimentally obtained values at 23 °C for solutes with relative molecular masses $M_r < 1000$ are smaller than the upper limit values obtained with equation 6 and $A_P = 11$. Taking into account the significant interaction between polyolefins and nonpolar solvents, for example olive oil at high temperatures, a corresponding T-dependence for A_P must be considered. The consequence is a higher apparent activation energy E_A (*9*). In the model of *Limm and Hollifield (1)*, the interaction between polyolefins and nonpolar solvents is also considered.

References

1. Limm, W.; Hollifield, H. C., *Food Additives and Contaminants* **1996,** *13,* 949.
2. Piringer, O., *Food Additives and Contaminants* **1994,** *11,* 221.
3. Baner, A. L.; Brandsch, J.; Franz, R.; Mercea, P.; Piringer, O., *Proceedings of the 17th Annual International Conference on Advances in the Stabilization and Degradation of Polymers:* Lucerne, Swiss, 1995, pp 11.
4. Baner, A. L.; Brandsch, J.; Franz, R.; Piringer, O., *Food Additives and Contaminants* **1996,** *13,* 587.
5. Marano, J. J.; Holder, G. D., *Ind. Eng. Chem. Res.* **1997,** *36,* 1887.
6. Piringer, O., *Verpackungen für Lebensmittel;* VCH Verlagsgesellschaft GmbH: Weinheim, New York, 1993, pp 246.
7. Reid, R. C.; Prausnitz, J., M.; Poling, B., E., *The Properties of Gases and Liquids;* McGraw-Hill Book Company, New York, 1987, pp. 656.
8. Koszinowski, J., *J. Applied polymer Science* **1986,** *31,* 1805.
9. Piringer, O., *Proceedings of the Pira Conference „Plastics for Packaging Food",* Prague, 1997.

Chapter 5

The Estimation of Partition Coefficients, Solubility Coefficients, and Permeability Coefficients for Organic Molecules in Polymers

Using Group Contribution Methods

A. L. Baner

Friskies R&D Center Inc., 3916 Pettis Road, St. Joseph, MO 64503

Partition, solubility and permeability coefficients of organic substances are necessary for modeling mass transfer phenomena (aroma permeation and scalping, polymer additive migration) in polymeric food packaging systems. The uncountable number of different polymer/organic molecule/food system combinations of interest coupled with the laborious and difficult experimental work needed for measurement makes it desirable to explore the use of semi-empirical thermodynamically based group contribution methods to estimate these parameters. The accuracy of partition, solubility and permeability coefficients estimations using the UNIFAC, GCFLORY, ELBRO-FV, Regular Solution and Retention Indices methods are compared with experimental data for aroma compounds and polymer additives in polyolefin, PET, nylon 6 and PVC polymers.

Partition coefficients, K, are fundamental physicochemical parameters describing the distribution of a solute between two contacting phases at equilibrium. Partition coefficient are directly related to the mole fraction activity coefficient of the solute, γ_i, in the two contacting phases of a system. Activity coefficient at very low solute concentrations, infinite dilution activity coefficient, γ_i^{∞}, are characteristic physical chemical system parameter. The infinite dilution activity coefficient characterizes the behavior of the solute molecule isolated in solvent where the solute molecule exhibits maximum non-ideality.

In food package systems, the concentrations of most flavor and aroma compounds and migrateable package components are present in the dilute concentration range. The dilute concentration range, which varies depending on the system under consideration, can be defined to be approximately a mole fraction of $1x10^{-5}$ which corresponds roughly to 100 ppm (w/v) for a typical aroma compound (MW $\approx$ 150) in an aqueous solvent system. While polymer engineers are interested in estimating the properties of polymer solutions, with polymers dissolved in low molecular weight solvents or mixtures of polymers, food engineers are concerned with mobile low molecular weight substances sorbed in a continuous "solid" polymer phase.

There has been significant effort devoted to activity coefficient estimation methods in the chemical engineering, environmental and pharmaceutical research fields because the necessary experimental data for many substances are not available and are difficult to measure.

The estimation of activity coefficients in polymer systems presents special problems for modeling and there has been much activity in recent years with the introduction of several activity coefficient estimation models. These models can be based on thermodynamic models with empirical corrections (1), equations of state (2, 3, 4, 5), statistical mechanics (6), quantum mechanical (7, 8) and free volume models (9).

One of the central problems with estimating activity coefficient in polymer systems is that general observations made for low molecular weight component systems are no longer valid for polymers. It is observed that the solution dependent properties are no longer directly proportional to the mole fraction of solute in the polymer at dilute concentrations. For example the solute partial pressure in a system containing a polymer is no longer directly proportional to its mole fraction which is an apparent deviation from Raoult's law.

Some estimation models work around the difficulty of using mole fractions for polymers by using weight fractions (molal concentration). This avoids the problems of defining what is a mole of polymer. However, even using weight fractions and defining the estimated activity coefficients on a molal basis is not enough to overcome the differences between what is expected from a low molecular weight liquid system and what is experimentally observed in a polymer system. This difference is treated in most estimation models by describing the difference as being due to free volume differences between the polymer and liquid (see for example Flory (2), UNIFAC-FV (6), GCFLORY (4, 5), ELBRO-FV(9)). The free volume concept was specifically developed to describe the variation of polymer system properties from those of a liquid system.

All current activity coefficient estimation models are by necessity semi-empirical in nature because still too little is known about solution theory for outright estimation. Chemical modeling is not readily available and is not far enough developed to do these types of calculations. The constants required by the models must be estimated using either experimental data points (e.g. an infinite dilution activity coefficient or a molar volume) or by using group contributions derived from experimental data (e.g. interaction constants, molecular volumes and surface areas).

The goal of this work is to take some of the more promising activity coefficient estimation models and apply them for estimating infinite dilution activity coefficient in both polymers and in liquids. The estimated activity coefficient can be used to derive partition coefficient and solubility coefficient for used in the modeling of mass transfer between polymeric packaging materials and food.

Definition of Activity Coefficient. Common thermodynamic notation for condensed phases (liquid (L) or polymer (P) mixtures) not described by an equation of state is to define an mole fraction activity coefficient γ_i (T, P, x_i) for a solute i, which is a function of temperature (T), pressure (P) and composition (x_i) by the equation (10):

$$f_i^L(T,P,x_i) = x_i \cdot \gamma_i(T,P,x_i) \cdot f_i^{\circ L}(T,P) \tag{1}$$

Where f is the fugacity and f° is the pure component standard state fugacity. The activity coefficient is related to the molar excess free energy of mixing by (10):

$$\sum n_i \cdot \overline{G_i}^{ex} = \sum n_i \cdot R \cdot T \cdot \ln\gamma_i(T,P,x_i) = \underline{G}^{ex} = \underline{H}^{ex} - T \cdot \underline{S}^{ex} \tag{2}$$

Where R is the gas constant, $\overline{G_i}^{ex}$ is the partial molar excess free energy of mixing, $\underline{G}^{ex}$ is the excess free energy of mixing per mole, $\underline{H}^{ex}$ is the excess enthalpy of mixing per mole and $\underline{S}^{ex}$ is the excess entropy of mixing per mole. Note that the excess free energy of mixing is also referred to as the excess chemical potential μ^{ex} in some notations. A regular solution is a special case where solutions of similar sized molecules are completely randomly oriented in solution (i.e. no attractive forces other than dispersion forces) so that the volume change on mixing is quite small and the excess entropy per mole of mixture is essentially zero. For regular solutions then $\underline{S}_i^{ex} = 0$ but $\underline{H}_i^{ex} \neq 0$. Another special case, athermal solutions, are assumed to have zero (or negligibly small) enthalpy of mixing.

The activity coefficient is commonly described in several of the estimation models as being roughly composed of two or three different components. These components represent combinatorial contributions (γ_i^c) which are essentially due to differences in size and shape of the molecules in the mixture, residual contributions (γ_i^r) which are essentially due to energy interactions between molecules, and free volume (γ_i^{fv}) contributions which take into consideration differences between the free volumes of the mixture's components:

$$\ln\gamma_i = \ln\gamma_i^c + \ln\gamma_i^r + \ln\gamma_i^{fv} \tag{3}$$

The combinatorial and free volume contributions can be thought of as being roughly analogous to the excess entropy of mixing. The residual contribution is roughly analogous to the excess enthalpy of mixing. In some models the free volume contribution is treated in a separate term (equation 3, e.g. UNIFAC and GCFLORY) and in others (ELBRO-FV) the combinatorial and free volume contributions are essentially combined into the free volume term thus eliminating the combinatorial term. The regular solution theory considers essentially only a residual contribution.

The concept of free volume varies on how it is defined and used but is generally acknowledged to be related to the degree of thermal expansion of the molecules. When liquids with different free volumes are mixed that difference contributes to the excess functions (11). The definition of free volume used by Bondi (12) is the difference between the hard sphere or hard core volume of the molecule ($\underline{V}^W$ = van der Waals volume) and the molar volume at some temperature:

$$\underline{V}^{fv} = \underline{V}_i - \underline{V}^W \tag{4}$$

Experimental Methods

All activity coefficient estimations using the above models were carried out using molar concentrations and weight fractions of 1×10^{-5} and a temperature of 25 °C. This corresponds roughly to the actual weight fractions and molar concentrations of solutes in the polymer and liquid phases (1×10^{-6} to 1×10^{-4}) of the experimental data. The experimental data used was collected by equilibrium sorption methods of a low concentration mixture of solutes in a solvent in contact with a polymer film (13, 14, 15) and permeation of solutes through a polymer film pouch immersed in a solvent bath (13,16, 17), or headspace equilibrium measurement (18) using the method of Kolb et al. (19). In all experiments the solutes were present at very dilute concentrations in the polymer and any solvents or solutes in contact with the polymer where sorbed in such small amounts so there were no significant changes in the partitioning or diffusion behaviors of the polymer.

Regular Solution Model. The regular solution theory (RST) which uses solubility parameters to estimate activity coefficients is one of the oldest and gives relatively good activity coefficients estimations for hydrocarbons. RST as its name implies is for regular solutions. It uses the geometric mean assumption which assumes interactions between different molecules in a mixture are similar to those the molecules experience between themselves in a pure mixture. Thus giving its best estimations for mixtures of similar sized nonpolar molecules. The use and applications of RST have been reviewed (20, 21). The accuracy of RST is not good for estimations of very dilute concentrations of organic compound partition coefficient between polyolefins and alcohols (*1*).

Activity coefficients were estimated using a correlation (*1*) using solubility parameters estimated by the group contribution method of Van Krevelen (*21*).

UNIFAC. The UNIFAC (**Uni**fied quasi chemical theory of liquid mixtures **F**unctional-group **A**ctivity **C**oefficients) group-contribution method for the prediction of activity coefficients in non-electrolyte liquid mixtures was first introduced by Fredenslund et al. (*22*). It is based on the unified quasi chemical theory of liquid mixtures (UNIQUAC) (*23*) which is a statistical mechanical treatment derived from the quasi chemical lattice model (*24*). UNIFAC has been extended to polymer solutions by Oishi and Prausnitz (*6*) who added a free volume contribution term (UNIFAC-FV) taken from the polymer solution equation-of-state equation of Flory (*2*). The UNIFAC activity coefficient estimation model uses the form of equation 3.

The UNIFAC method has a useful temperature range of -23 to 152 °C. The accuracy of infinite dilution estimations for relatively low molecular weight organic molecules ($M_i < 200$) averages from 20.5% for 3357 low molecular weight compounds (*25*), 21% for 6 compounds in three classes of solvents (*26*) and 21.1 % for 791 series of measurements containing 1773 data points *(27)*. Several modifications have been proposed to the UNIFAC model in addition to the free volume contribution already mentioned. Proposed modifications to the combinatorial and residual terms that improve estimations by about 11% (*27, 28*) but cannot be used for polymers (*29*). Other UNIFAC modifications have been developed (*30, 31, 32*) that attempt to increase the predictions made by the original model for infinite dilutions, hydrocarbon solubility and swelling/dissolving of semicrystalline polymers.

Mixtures of hydrocarbons are assumed to be athermal by UNIFAC meaning there is no residual contribution to the activity coefficient. The free volume contribution is considered significant only for mixtures containing polymers and is equal to zero for liquid mixtures. The combinatorial activity coefficient contribution is calculated from volume and surface area fractions of the molecule or polymer segment. The molecule structural parameters needed to do this are the van der Waals or hard core volumes and surface areas of the molecule relative to those of a standardized polyethylene methylene CH_2 segment. UNIFAC for polymers (*6*) calculates in terms of activity (a_i) instead of the activity coefficient and uses weight fractions ($w_i = g_i \Big/ \sum_{j=1}^{j} g_i$) instead of mole fractions where g_i is the mass of component i and there are j components in the mixture.

UNIFAC estimations of activity coefficients were estimated using a BASIC computer program based on a program (*10*) modified for calculations for polymers (UNIFAC-FV) to carry out the free volume correction using weight fraction based activity coefficients (*6*). The program was rewritten using published algorithms (*33*) for binary polymer/solute solutions and the interaction parameters were updated using the UNIFAC 5th revision interaction parameters (*34*). In UNIFAC calculations the polymer monomer repeat unit is used to represent the chemical structure and molecular weight of the polymer in the polymer activity coefficient calculations. For polyethylene (PE) polymers the amorphous PE density ($\rho_P = 0.85$) gave the most accurate estimations.

Group-Contribution Flory Equation-Of-State. The group-contribution Flory equation-of-state (GCFLORY) developed by Chen et al. (*4*) and later revised (*5*) is a group-contribution extension of the Flory equation of state (*2*). The equation is similar to the Holten-Anderson model (*35*) and incorporates a correlation for the degree of freedom parameter. The method has combinatorial and free volume contributions to the activity coefficient similar to UNIFAC and uses the UNIFAC surface area and volume functional group parameters. The model estimates activities and uses weight fractions.

Copies of the FORTRAN versions of the GCFLORY programs were obtained from the authors (POLGCEOS March 5, 1991, Chen et al. 1990 and GC-FLORY EOS April 28, 1993, Bogdanic and Fredenslund, 1994). An average number molecular weight of 30,000 was used in the calculations for all polymers. The model's resulting activity coefficient estimations are relatively insensitive to variations in polymer molecular weights.

Elbro Free Volume Model. The Elbro free volume model (ELBRO-FV) (*36*) is based on the free volume term proposed by Elbro et al. (*37*). The ELBRO-FV model uses only the free volume and residual activity coefficient contribution terms. The residual term is taken from the UNIFAC model and is thus equal to zero for athermal mixtures. For the activity coefficient estimation calculations the residual portion of the activity coefficient (Equation 3) was calculated in a simple spreadsheet and added to the interaction activity coefficient portion calculated using UNIFAC. A number average molecular weight of 30000 was used for the polymers. The van der Waals volumes were calculated using the group contribution method of Bondi (*38*).

Polymer/Liquid Partition Coefficient Equations.

By definition at equilibrium the fugacities of contacting phases are equal (i.e. $f_i^G = f_i^L = f_i^P$). For dilute concentrations (w/v) of solute in polymer and liquid phases, the respective mole fractions can be approximated by the following equations:

$$x_i^P = \frac{c_i^P \cdot \underline{V}_P}{M_i} \tag{5}$$

$$x_i^L = \frac{c_i^L \cdot V_L}{M_i} \tag{6}$$

Where M_i is the solute molecular weight, $\underline{V}_L$ is the molar volume of the liquid and $\underline{V}_P$ is the molar volume of the polymer. At equilibrium the concentration partition coefficient ($K_{P/L}$) (governed by the Nernst distribution law) can be defined as the ratio of the concentration (w/v) of the solute in the polymer (c_i^P) to the concentration (w/v) of the solute in the liquid (c_i^L). Combining equations 5 and 6 with the polymer-liquid equilibrium condition between the polymer and liquid gives:

$$K_{P/L} = \frac{c_i^P}{c_i^L} = \frac{\gamma_i^L \cdot \underline{V}_L}{\gamma_i^P \cdot \underline{V}_P} \tag{7}$$

The polymer/liquid coefficient can also be calculated from the ratio of the polymer/gas ($K_{P/G}$) and liquid/gas partition ($K_{L/G}$) coefficients:

$$K_{P/L} = \frac{K_{P/G}}{K_{L/G}} = \frac{c_P/c_G}{c_L/c_G} \tag{8}$$

For systems involving polymers, weight fractions (w_i) are often used instead of mole fractions thus defining a weight fraction (molal) basis activity coefficient (e.g. for liquids $\Omega_i^L = \left(M_L/M_i\right) \cdot \gamma_i^L$) at dilute concentrations which by convention is:

$$\Omega_i = \frac{a_i}{w_i} \tag{9}$$

Dilute concentrations, e.g. concentration of i in a polymer, c_i^P, can be approximated using weight fractions as:

$$c_i^P \cong w_i \cdot \rho_P = \frac{a_i}{\Omega_i^P} \rho_P \tag{10}$$

where a_i is the activity. Combining Equations 9 and 10 gives the equilibrium concentration partition coefficient in terms of weight fraction activity coefficients:

$$K_{P/L} = \frac{c_i^P}{c_i^L} = \frac{\Omega_i^L \cdot \rho_P}{\Omega_i^P \cdot \rho_L} \tag{11}$$

Using the solute's mole fraction activity coefficient in a liquid solvent along with a molal activity coefficient for the polymer phase one gets by combining equations 6, 7, and 10:

$$K_{P/L} = \frac{\gamma_i^L \cdot V_L \cdot \rho_P}{\Omega_i^P \cdot M_i} \tag{12}$$

Assuming an ideal gas phase and atmospheric pressure ($c_i^G = \dfrac{n_i \cdot M_i}{V} = \dfrac{f_i \cdot M_i}{R \cdot T}$), the liquid gas partition coefficient for a dilute solution can be estimated using mole fraction activity coefficient by combining the ideal gas phase expression with equations 1 and 6:

$$K_{L/G} = \frac{c_i^L}{c_i^G} = \frac{V_L}{x_i} \cdot \frac{R \cdot T}{f_i^\circ} = \frac{R \cdot T}{\gamma_i^L \cdot f_i^\circ \cdot V_L} \tag{13}$$

Similarly, an expression for a partition coefficient between polymer and gas from a molal activity coefficient in a dilute solution at atmospheric pressure can be derived by assuming an ideal gas and by using equation 10 and the definition of activity ($a_i = f_i/f_i^\circ$):

$$K_{P/G} = \frac{c_i^P}{c_i^G} = \frac{R \cdot T}{f_i^\circ} \cdot - = \frac{R \cdot T \cdot \rho_P}{\Omega_i^L \cdot M_i \cdot f_i^\circ} \tag{14}$$

The assumptions of dilute concentrations and solutes behaving as ideal gases at normal temperatures and pressures in equations 13 and 14 introduces an error of only a few percent which is easily within the uncertainty of most experimental measurements.

Estimation of solubility coefficients. The partition coefficient describing solute partitioning between air and polymer is often referred to as a solubility coefficient (S) (36). The solubility coefficient can be expressed in terms of weight fraction activity coefficients for polymers using equation 10:

$$S = \frac{c_i^P}{P_i} = \frac{\rho_P}{\Omega_i^P \cdot P_i^\circ} \tag{15}$$

with cgs units of $\dfrac{g}{cm^3 \cdot Pa}$ and where P_i is the solute partial pressure and P_i° is the solute saturated vapor pressure at the temperature of the system. Note that at ambient temperature and pressure the fugacity (f) is essentially the partial pressure (P). Henrys constant (H) is a special solubility coefficient case and can be expressed in the form: $c_i^P \propto H \cdot P_i^\circ$.

For n-alkanes with M_i > hexadecane, where no vapor pressure data exists (solid at 25 °C), the vapor pressures of the n-alkane homologous series was extrapolated for the larger molecules. For aroma compounds where no vapor pressure data exists they were estimated using vapor pressure data of substances with similar structures and retention indices.

Estimation of permeability Coefficient. The permeability coefficient (P) can be calculated by multiplying the estimated solubility coefficient (equation 15) and an estimated diffusion coefficient of the solute in the polymer (D_P):

$$P = D_P \cdot S \tag{16}$$

with cgs units for the permeability coefficient of ($\dfrac{g \cdot cm}{cm^2 \cdot Pa \cdot s}$). Note that equation 16 is for dilute concentrations of solutes in polymers and may not hold true for organic solutes over a wide concentration range. The diffusion coefficient in different polymers were estimated using estimation methods (*39, 40*) which correlate the diffusion coefficient with the type of polymer, solute molecular weight and temperature.

Results and Discussion

Partition Coefficient. A test set of 6 to 13 aroma compound partition coefficient between different food contact polymers (low density polyethylene (LDPE), high density polyethylene (HDPE); polypropylene (PP), polyethylene terephthalate (PET), poly amide (PA)) and different food simulant phases (water, ethanol, aqueous ethanol/water mixtures, methanol, 1-propanol) were taken from the literature (*13, 14, 15, 16, 17, 18, 45*). Table I shows the test set of 13 different aroma compounds used, their properties and their structures. The experimental data were compared to estimations using the different estimation methods described above.

Table II, with partition coefficient estimation results for 13 aroma compounds partitioned between polyethylene (PE) and ethanol, shows an example of the estimation accuracy one can expect using these methods. In order to compare the different estimation methods, average absolute ratios of calculated to experimental values were calculated partitioned substances. When the calculated values are greater than experimental values the calculated value is divided by the

Table I. Aroma compounds studied

Aroma compound	Molecular weight (g/mol)	Experimental molar volume 25 °C (mL/mol)	van der Waals molar volume (Bondi, 1968) (mL/mol)	Saturated Vapor Pressure 25 °C (Pa)	Structure
d-limonene	136.24	161.98	88.35	272.5[1]	
diphenylmethane	168.23	168.1	101.93	4.61[1]	
linalylacetate	196.23	219.32	124.08	14.2[1]	
camphor	152.23	153.27	96.8	41.5[1]	
diphenyloxide	170.21	158.99	96.7	2.69[2]	
isoamylacetate	130.18	150.39	83.47	723[1]	
undelactone	184.28	194.18	117.51	0.457[2]	
eugenol	164.2	153.98	98.98	2.63[1]	
citronellol	156.27	182.77	110.23	5.86[1]	
dimethylbenzyl-carbinol (DMC)	150.22	154.5	94.72	17.4[2]	
menthol	156.27	173.63	106.9	16.1[1]	
phenylethylalcohol (PEA)	122.17	120.36	74.31	11.2[1]	
cis-3-hexenol	100.16	118.49	69.26	140[1]	

1) Experimental data from (42) fitted using Miller equation (43)
2) Estimated using substances with similar retention indices.

Table II. Partition coefficients of 13 aroma compounds between 100 % ethanol and polyethylene

Aroma compound	Exper.[1] $K_{P/L}$	UNIFAC-FV	GCFLORY (1990)	GCFLORY (1993)	ELBRO-FV	UNIFAC (L) GCFLORY (P) (1990)	UNIFAC (L) GCFLORY (P) (1993)	UNIFAC (L) ELBRO-FV (P)	Regular Solution Theory	Retention Indices[2]
d-limonene	0.67	2.18	4.3	82	0.92	0.37	0.22	1.9	0.3	0.24
diphenylmethane	0.27	1.09	0.040	36	0.27	0.096	1.4	1.2	0.17	0.22
linalylacetate	0.058	0.31	1.5	20	0.15	0.18	0.60	0.34	0.074	0.075
camphor	0.062	0.36	0.029	2.3	0.21	0.012	0.097	0.38	0.16	0.031
diphenyloxide	0.23	0.68	0.52	13	0.20	0.48	0.48	0.75	0.027	0.22
isoamylacetate	0.0464	0.27	0.30	1.6	0.16	0.26	0.41	0.26	0.053	0.021
undelactone	0.052	0.66	0.20	3.4	0.32	0.20	0.32	0.63	-	0.052
eugenol	0.0181	0.00093	0.0051	0.034	0.00051	0.0019	0.0041	0.0010	0.008	0.015
citronellol	0.025	0.018	0.046	0.39	0.019	0.052	0.15	0.22	-	-
DMC	0.0113	0.0091	-	0.22	0.0082	-	0.12	0.012	-	0.0095
menthol	0.02	0.019	0.0088	0.24	0.017	0.0097	0.081	0.23	0.027	0.011
PEA	0.0097	0.0067	0.0037	0.13	0.0076	0.010	0.11	0.010	0.0029	0.0019
cis-3-hexenol	0.0139	0.014	0.034	0.16	0.015	0.040	0.15	0.016	0.0029	0.022
(undelactone)	0.052	-	-	2.8	-	-	0.47	-	-	-
(limonene)	0.67	-	0.011	64	-	1.2	0.60	-	-	-
(camphor)	0.062	-	0.25	1.8	-	0.23	0.21	-	-	-
(menthol)	0.02	-	0.012	0.17	-	0.017	0.11	-	-	-
AAR		5.0	5.6 (11)	67 (63)	4.7	3.5 (3.2)	6.5 (6.8)	4.89	2.91	2.27
AAR s.d.		5.2	6.4 (17)	90 (89)	9.0	2.2 (2.3)	3.3 (3.4)	4.77	2.15	1.64
AAR c.v. %		103	115 (156)	134 (141)	193	63 (70)	52 (50)	97.5	73.9	72.5

1) Experimental data from (13, 14). 2) Calculated using method from (44).
( ) calculated using cyclic group contribution parameters (GCFLORY only)
Calculated using polymer density = 0.85., (L) = liquid phase activity coefficient, (P) = polymer phase activity coefficient, - not possible to calculate a result
AAR = average absolute ratio: for calculated (calc.) > experimental (esp.) = calc./exp.; for calc. < exp. = exp./calc.
AAR s.d. = standard deviation of absolute ratio. AAR c.v. % = % coefficient of variation absolute ratio.

experimental value. For calculated values less than the experimental values the inverse ratio is taken. Calculating absolute ratios gives a multiplicative factor indicating the relative differences between values of the experimental and estimated data. A ratio of one means the experimental value is equal to the estimated value.

The liquid phase and polymer phase activity coefficients were combined from different methods to see if better estimation accuracy could be obtained since some estimation methods were developed for estimation of activity coefficients in polymers (e.g. GCFLORY, ELBRO-FV) and others have their origins in liquid phase activity coefficient estimation (e.g. UNIFAC). The UNIFAC liquid phase activity coefficient combined with GCFLORY (1990 and 1993 versions) and ELBRO-FV polymer activity coefficients were shown to be the combinations giving the best estimations out of all the possible combinations of the different methods. Also included in Table II are estimations of partition coefficients made using a semi-empirical group contribution method referred to here as the "Retention Indices Method". This method as its name implies is based on the concept of retention indices from gas chromatography. So far the method only has group contribution units worked out for the estimating the partitioning of organic substances between polyolefins in contact with ethanol and water but gives very good estimations for these systems.

Table III shows examples of average absolute ratios (estimated values to experimental values) and their corresponding standard deviations for several polymer/aroma/solvent systems. This is representative of the estimation behavior of these models and does not include all data tested to date. UNIFAC-FV is the most consistent, most widely applicable and overall gives the most accurate partition coefficient estimations of all the models. However, the UNIFAC model partition coefficient estimations here are less accurate than the 20 % variation reported in the literature (*25, 26*) for solute in solvent activity coefficient data. GCFLORY (1990) could not estimate water and polar polymers (PVC, PET, no amide group for polyamide) containing systems well. GCFLORY (1993) is very similar to GCFLORY (1990) but made worse estimations because it was not intended to be used for estimating activity coefficients in low molecular weight liquids. When the GCFLORY polymer activity coefficient was used with the liquid phase activity coefficient from UNIFAC, partition coefficient estimations were significantly improved. The GCFLORY models had cyclic and aliphatic group contribution terms. Calculations were made for 4 ring containing aroma compounds using both cyclic and aliphatic groups to see if there was any advantage of one set of groups over the other for these estimations. In most cases the results were very similar either slightly better or slightly worse.

ELBRO-FV cannot model water as a solvent otherwise it was often better than UNIFAC-FV in accuracy. For water containing systems UNIFAC liquid activity coefficient estimations should be used with ELBRO-FV polymer activity coefficients. The scope of application of the Regular Solution Theory and Retention Indices is limited only to solute partitioning in PE/ethanol systems. Their better

Table III. Average absolute ratios of estimated aroma compound partition coefficients between polymers and solvents

Partition System	UNIFAC-FV	GCFLORY (1990)	GCFLORY (1993)	ELBRO-FV	UNIFAC (L) GCFLORY (P) (1990)	UNIFAC (L) GCFLORY (P) (1993)	UNIFAC (L) ELBRO-FV (P)	Regular Solution Theory	Retention Indices
PE (ρ_P = 0.918) /13 aromas/ 100 % Ethanol[1]	5.0 (5.2)	5.6 (6.4) c:11 (17)	67 (90) c:(63) (89)	4.7 (9.0)	3.5 (2.2) c:3.2 (2.3)	6.5 (3.3) c:6.8 (3.4)	4.89 (4.8)	2.9 (2.2)	2.3 (1.6)
PE (ρ_P = 0.918) /13 aromas/ 100 % water[2]	17.7 (30.3)	8060 (24300) c:2120 (607)	653000 1.9E+6 (90) c:653000 (1.9E+6)	4.2E+19 (1.1E+20)	10.2 (10.0) (6.2) (3.9)	45 (75) c:43 (75)	18 (29)	-	-
PP (ρ_P = 0.902) /13 aromas/100 % Ethanol	5.8 (6.8)	9.2 (15) c:11 (21)	133 (147) c:115 (130)	5.7 (12)	5.5 (6.6) c:4.6 (6.3)	9.9 (6.3) c:11 (6.8)	5.8 (6.6)	-	-
PVC (ρ_P = 1.36) /13 aromas/100 % Ethanol[2]	14 (27)	59 (122) c:62 (122)	73 (181) c:68 (178)	24 (53)	122 (334) c:131 (347)	16 (21) c:14 (17)	13 (26)	-	-
PET (ρ_P = 1.37) /7 aromas/100 % Methanol[3]	2.9 (1.8)	1380 (2210) c:922 (2240)	2.6E+6 (3.2E+6) c:3.6E+6 (4.3E+6)	4.4 (3.1)	620 (1220) c:4.8 (4.6)	3.3 (3.3) c:3.4 (3.2)	4.1 (3.8)	-	-
PET (ρ_P = 1.37) /7 aromas/100 % 1-Propanol[3]	12 (11)	73 (85) c:27 (49)	127 (199) c:127 (199)	10 (12)	250 (480) 4.6 (4.0)	15 (18) 15 (18)	18 (22)	-	-
PA (ρ_P = 1.14) /6 aromas/100 % Methanol[3]	4.5 (2.9)	-	-	7.9 (3.5)	-	-	6.6 (5.0)	-	-
PA (ρ_P = 1.14) /7 aromas/100 % 1-Propanol[4]	6.1 (2.7)	-	-	7.0 (3.8)	-	-	12 (8.2)	-	-
Average AAR:	8.5	1598	100	9.1	169	16	10.3	-	-

1) Experimental data from (13, 14), 2) Experimental data from (13), 3) Experimental data from (15), 4) Experimental data from (17)

c calculated using cyclic group contribution parameters (GCFLORY only).   - not possible to calculate a result

AAR = average absolute ratio: for calculated (calc.) > experimental (esp.) = calc./exp.; for calc. < exp. = exp./calc.

( ) = AAR s.d. = standard deviation of absolute ratio.

accuracy is not surprising since they are essentially correlations of this experimental data.

Solubility and Permeability Coefficients. The lack of sufficient experimental data has restricted the testing of estimated solubility and permeability coefficients to polyolefins. Table IV shows typical estimated values for solubility and permeability coefficients of thirteen aroma compounds in PP. The lower accuracy of estimated permeability coefficient values compared to estimated solubility coefficients is due to additional errors from using estimated diffusion coefficients whose values tend to be too large. The average absolute ratios and standard deviations of the absolute ratios of the diffusion coefficient estimations to experimental values for aromas in the different polyolefins were 3.8 ± 2.9 for LDPE, 3.1 ± 1.5 for HDPE and 9.7 ± 9.3 for PP. For n-alkanes the average absolute ratios were 5.3 ± 1.1 for LDPE and 7.0 ± 1.9 for HDPE. The development of highly accurate diffusion coefficient models for all types of substances in polymers will continue to be a limiting factor in the accuracy of permeability coefficients. Additional error in these estimations comes from the accuracy of the solute vapor pressure which is necessary to calculate the solubility coefficient.

Table V summarizes the results for estimation of solubility and permeability coefficients for 13 aromas and 12 n-alkanes (C5 - C10, C12, C14, C16, C18, C20, C22) in the polyolefins. All estimation models gave similar estimation accuracy's. However, like for the partition coefficient the UNIFAC-FV was the most accurate followed closely by ELBRO-FV and then GCFLORY (both models were similar). The Regular Solution Theory, which can only be used for n-alkanes, gave the worst estimation of all the models. This confirms the established fact that the Regular Solution Theory is a qualitative method for indicating relative trends and tendencies in polymer solubility.

In all models occasionally an anomalous estimation result would appear. The user of these models must be vigilant and critical of the model outputs. Partition coefficient for substances with similar polarities in a given polymer solvent system will not vary greatly from one another. As a point of reference very large partition coefficient $> 1 \times 10^5$ practically exist only for non-polar alkanes partitioned between polyolefins and water. Similarly, n-Alkanes in polyolefins represent the upper limit for solubility coefficients for molecules with similar molecular weights. Other upper limit can be found for substances that swell or dissolve PET, PVC and PA polymers.

Conclusions

Semi-empirical thermodynamic based group contribution models can be used to estimate partition, solubility and permeability coefficients. Ways to estimate these parameters using these models have been outlined and results calculated. These models allow the estimation of activity coefficients based solely on the molecular structures of the system's components.

Table IV. Solubility and Permeability Coefficients of 13 aroma compounds in and through polypropylene (PP)

Aroma compound	Exper. S $\frac{g}{cm^3 \cdot Pa}$	Exper. P $\frac{g \cdot cm}{cm^2 \cdot Pa \cdot s} \times 10^{14}$	UNIFAC-FV S_{est} $\frac{g}{cm^3 \cdot Pa}$	UNIFAC-FV P_{est} $\frac{g \cdot cm}{cm^2 \cdot Pa \cdot s} \times 10^{14}$	GCFLORY (1990) S_{est} $\frac{g}{cm^3 \cdot Pa}$	GCFLORY (1990) P_{est} $\frac{g \cdot cm}{cm^2 \cdot Pa \cdot s} \times 10^{14}$	GCFLORY (1993) S_{est} $\frac{g}{cm^3 \cdot Pa}$	GCFLORY (1993) P_{est} $\frac{g \cdot cm}{cm^2 \cdot Pa \cdot s} \times 10^{14}$	ELBRO-FV S_{est} $\frac{g}{cm^3 \cdot Pa}$	ELBRO-FV P_{est} $\frac{g \cdot cm}{cm^2 \cdot Pa \cdot s} \times 10^{14}$
d-limonene	0.00037	1.18	0.000857	13	0.000363	5.5	0.000219	3.3	0.000777	12
diphenylmethane	0.0036	5.76	0.0297	245	0.0184	152	0.0318	263	0.0336	277
linalylacetate	0.0028	2.18	0.00556	28	0.00139	6.9	0.0137	68	0.000764	38
camphor	0.000094	7.43	0.00387	43	0.0106	118	0.00138	15	0.00367	41
diphenyloxide	0.00093	1.86	0.0396	315	0.00656	52	0.0264	210	0.0430	343
isoamylacetate	0.00038	1.44	0.000138	2.4	0.000153	2.6	0.000241	4.1	0.000152	2.6
undelactone	--	--	--	--	--	--	--	--	--	--
eugenol	0.0014	2.1	0.00189	17	0.00410	37	--	--	0.00189	17
citronellol	0.001	0.71	0.00504	52	0.00318	33	0.0306	316	0.00509	53
DMC	0.001	1.1	0.00105	12	0.00470	49	0.00659	76	0.000958	11
menthol	0.00024	0.062	0.00209	22	0.000976	49	0.00758	78	0.00192	20
PEA	0.00032	0.544	0.00111	23	0.000206	20	0.00772	156	0.00109	22
cis-3-hexenol	0.000075	1.13	0.000133	4.3	--	6.7	0.000916	30	0.000128	4.1
(undelactone)	--	--	--	--	--	--	--	--	--	--
(limonene)	0.00037	1.18	--	--	0.000273	4.2	0.000463	7.1	--	--
(camphor)	0.000094	7.43	--	--	0.0103	115	0.00251	28	--	--
(menthol)	0.00024	0.062	--	--	0.00293	30	0.00886	92	--	--
AAR			10	61	14 (13)	84 (60)	15 (17)	207 (227)	10	60
AAR s.d.			14	97	30 (29)	210 (129)	11 (12)	357 (414)	15	92
AAR c.v. %			144	160	224 (232)	250 (218)	74 (75)	172 (182)	145	156

1) Experimental data from (*36*). Permeability coefficient calculated from S experimental times D experimental. Polymer density $\rho_P = 0.902$.

2) S_{est} = Solubility coefficient estimated

3) P_{est} = S_{est} x D_{est} = estimated permeability coefficient

( ) calculated using cyclic group contribution parameters (GCFLORY only)

- not possible to calculate a result

AAR = average absolute ratio: for calculated (calc.) > experimental (exp.) = calc./exp.; for calc. < exp. = exp./calc.

AAR s.d. = standard deviation of absolute ratio. AAR c.v. % = % coefficient of variation absolute ratio.

Table V. Average absolute ratios of experimental to calculated solubility and permeability coefficients

Test System	UNIFAC-FV S_{est}	UNIFAC-FV P_{est}	GCFLORY (1990) S_{est}	GCFLORY (1990) P_{est}	GCFLORY (1993) S_{est}	GCFLORY (1993) P_{est}	ELBRO-FV S_{est}	ELBRO-FV P_{est}	Regular Solution Theory S_{est}	Regular Solution Theory P_{est}
	AAR ± s.d.	AAR ± s.d.	AAR ± s.d.	AAR ± s.d..	AAR ± s.d.	AAR ± s.d.	AAR ± s.d.	AAR ± s.d.	AAR ± s.d.	AAR ± s.d.
13 aromas LDPE[1] ($\rho_P = 0.918$)	9.3 ± 14	21 ± 23	8.0 ± 11	17 ± 18	14 ± 13	53 ± 89	10 ± 16	25 ± 26	-	-
c	-		8.4 ± 11	22 ± 19	15 ± 13	89 ± 90	-	-	-	-
13 aromas HDPE[1] ($\rho_P = 0.956$)	8.1 ± 12	10 ± 20	10 ± 8.9	6.9 ± 23	10 ± 8.9	15 ± 16	8.9 ± 13	11 ± 22	-	-
c	-	-	10 ± 8.7	11 ± 17	10 ± 8.9	16 ± 25	-	-	-	-
13 aromas PP[1] ($\rho_P = 0.902$)	10 ± 14	61 ± 97	14 ± 30	84 ± 210	15 ± 11	207 ± 357	10 ± 15	60 ± 92	-	-
c	-	-	13 ± 29	60 ± 129	17 ± 12	227 ± 414	-	-	-	-
12 n-alkanes LDPE[2] ($\rho_P = 0.918$)	5.1 ± 2.5	19 ± 18	6.3 ± 2.9	24 ± 22	6.3 ± 2.9	24 ± 22	7.2 ± 3.0	26 ± 24	84 ± 53	336 ± 336
12 n-alkanes HDPE[2] ($\rho_P = 0.956$)	1.7 ± 0.58	8.0 ± 8.7	3.3 ± 1.1	16 ± 17	3.3 ± 1.1	16 ± 17	3.7 ± 1.1	17 ± 18	42 ± 22	225 ± 259
Average AAR:	6.8	23.8	8.3	30	9.7	63	8.0	27.8	-	-

1) Experimental data from (36). 2) Experimental permeation data (16) experimental solubility data (18)

S_{est} = Solubility coefficient estimated.

P_{est} = S_{est} × D_{est} = estimated permeability coefficient

c calculated using cyclic group contribution parameters (GCFLORY only). - not possible to calculate a result

AAR = average absolute ratio: for calculated (calc.) > experimental (esp.) = calc./exp.; for calc. < exp. = exp./calc.

AAR s.d. = standard deviation of absolute ratio.

Based on this set of test data and other published work (*1, 14, 41*), which represents a cross section of the important food contact polymers and a series of typical aroma compound structures, UNIFAC-FV would be recommended over the other estimation models because of its availability, accuracy, and broad range of application. UNIFAC is widely used and continuously being improved as evidenced by the large amount of literature published on it. Although UNIFAC has some inherent weaknesses, i.e. it is based on regression of available experimental data and there are assumptions that may limit its accuracy for certain systems; the model works best in this application for most systems. Combining different models may lead to increased estimation accuracy for some systems but the extra effort to do this is not justified by the small increase in accuracy.

The lack of enough good experimental data will limit the testing of these estimation models in other polymers besides the polyolefins.

In general partition coefficient estimations tend to be less accurate (e.g. factor 3 to 18 for these systems) because estimations require activity coefficient estimations in two phases (polymer and liquid) both of whose errors can be compounded in the final estimation. Solubility coefficient estimations tend to be more accurate (e.g. factor 1.7 to 10 for these systems) since only the polymer phase activity coefficient estimations are required. A complicating factor in estimating solubility coefficients is the availability of accurate vapor pressures for the substances being estimated (especially non volatile substances e.g. polymer additives). The accuracy of the permeability coefficient is the lowest of the three parameters (e.g. factor range 8 to 61). The accuracy of the permeability coefficient is affected by the compounding of errors from the estimated solubility and diffusion coefficients. Improving the accuracy of the estimated diffusion coefficients for polymers other than polyolefins will be key in improving permeability coefficient estimation accuracy.

Acknowledgments
In addition to current support of Nestec S.A., parts of this work have been carried out in previous years at Fraunhofer-Institut für Lebensmitteltechnologie und Verpackung ILV, Munich, Germany and the Michigan State University School of Packaging Center for Food and Pharmaceutical Packaging Research.

Literature Cited

1. Baner, A.L., Piringer, O. *Ind. Eng. Chem. Res.* **1991**, *30*: 1506-1515.
2. Flory, P.J. *Discuss. Faraday Soc.*, **1970**, *48(7):* 7-29.
3. High, M.S., Danner, R.P. *Fluid Phase Equilibria*, **1990**, *55*: 1-15.
4. Chen, F., Fredenslund, A., Rasmussen, P. *Ind. Eng. Chem. Res.*, **1990**, *29*: 875-882.
5. Bogdanic, G., Fredenslund, A. *Ind. Eng. Chem. Res.* **1994**. *33*, 1331-1340.

6. Oishi, T., Prausnitz, J.M. *Ind. Eng. Chem. Process Des. Dev.*, **1978**, *17(3):* 333-339.

7. Mavrovouniotis, M.L. *Ind. Eng. Chem. Res.* **1990**. *29*, 1943-1953.

8. Canstantinou, L.; Prickett, S.E.; Mavrovouniotis, M.L. *Ind. Eng. Chem. Res.* **1993**, *32*, 1734-1746.

9. Kontogeorgis, G.M.; Fredenslund, A.; Tassios, D.P. *Ind. Eng. Chem. Res.*, **1993**, *32*, 362-372.

10. Sandler, S.I. *Chemical and Engineering Thermodynamics*. John Wiley and Sons; New York, N.Y., 1989. Chapter 7; pp 321.

11. Prausnitz, J.M.; Lichtenthaler, R.N.; Gomes de Azevedo, E.; Molecular *Thermodynamics of Fluid-Phase Equilibria*, Prentice-Hall: Englewood Cliffs, N.J., 1986; Chapter 6, pp 193.

12. Bondi, A. *Physical properties of Molecular Crystals, Liquids, and Glasses*. John Wiley and Sons, New York, 1968. Chapter 8.7.1, pp 255.

13. K., Koszinowski, J., Piringer, O. *Verpackungs-Rundschau*. **1989**, *40(5):* 39-44.

14. Baner, A.L., *Partition coefficients of aroma compounds between polyethylene and aqueous ethanol and their estimation using UNIFAC and GCFEOS*. Ph.D. Dissertation, Michigan State University, E. Lansing, 1992.

15. Franz. R. Personnal communication. Fraunhofer-Institut, Munich, 1990.

16. Koszinowski, J. *J. Appl. Poly. Sci.* **1986**, *31*, 1805-1826.

17. Franz, R. Fraunhofer-Institut für Lebensmitteltechnologie und Verpackung ILV *Annual Report* 1991. Munich, pp 47.

18. Baner, A.L. unpublished data. 1993.

19. Kolb, B; Welter, C.; Bichler, C. *Chromatographia*, **1992**, *34*, 235-240.

20. Barton, A.F.M. CRC *Handbook of Solubility Parameters and Other Cohesion Parameters*. CRC Press, Inc. Boca Raton, Florida, 1983.

21. Van Krevelen, D.W. *Properties of Polymers: Their Correlation With Chemical Structure; Their Numerical Estimation and Prediction From Additive Group Contributions*. 3rd edition Elsevier Scientific Pub. Co.: Amsterdam, 1990. Chapter 7, pp 189.

22. Fredenslund, A., Jones, R.L., Prausnitz, J.M. *AICHe Journal*, **1975**, *21(6):* 1086-1099.

23. Abrams, D.S., Prausnitz, J.M. *AICHe Journal*. 1975, **21**, 116-128.

24. Guggenheim, E.A. *Mixtures*. Clarendon Press, Oxford, 1952.

25. Thomas, E.R., Eckert, C.A. *Ind. Eng. Chem. Process Des. Dev.*, **1984**, *23:* 194-209.

26. Park, J.H., Carr, P.W. *Anal. Chem.*, **1987**, *59:* 2596-2602.

27. Weidlich, U., Gmehling, J. *Ind. Eng. Chem. Res.* **1987**. *26:* 1372-1381.

28. Larsen, B.L., Rasmussen, P., Fredenslund, A. *Ind. Eng. Chem. Res.*, **1987**, *26:* 2274-2286.

29. Fredenslund, A. *Fluid Phase Equilibria*, **1989**, *52:* 135-150.

30. Bastos, J.C., Soares, M.E., Medina, A.G. *Ind. Eng. Chem. Res.* **1988**, *27:* 1269-1277.

31. Iwai, Y., Ohzono, M., Aria, Y. *Chem. Eng. Commun.*, **1985**, *34*: 225-240.

32. Doong, S.J., Ho, W.S.W. *Ind. Eng. Chem. Res.* **1991**, *30*: 1351-1361.

33. Goydan, R., Reid, R.C., Tseng, H. *Ind. Eng. Chem. Res.*, **1989**, *28*: 445-454.

34. Hansen, H.K., Rasmussen, P., Fredenslund, A., Schiller, M., Gmehling, J. *Ind. Eng. Chem. Res.* **1991**, *30*: 2352-2355.

35. Holten-Anderson, J., Rasmussen, P., Fredenslund, A. *Ind. Eng. Chem. Res.*, **1987**, *26*: 1382-1390.

36. Becker, K., Koszinowski, J., Piringer, O. *Deutsche Lebensmittel-Rundschau.* **1983**, *79(8):* 257-266.

37. Elbro, H.S., Fredenslund, A., Rasmussen, P. *Macromolecules*, **1990**, *23 (21)*, 4707.

38. Bondi, A. *Physical properties of Molecular Crystals, Liquids, and Glasses.* John Wiley and Sons, New York, 1968. Chapter 14, pp 450.

39. Baner. A.L.; Brandsch, J.; Franz, R.; Piringer, O. *Fd. Addit. Contam.* **1996**, *13*, 587-601.

40. Brandsch, J.; Mercea, P.; Piringer, O. *New Developments in the Chemistry of Packaging Materials.* ACS, Washington, D.C. 1999.

41. Baner, A.L.; Franz, R.; Piringer, O. *Food Packaging and Preservation.* Blackie Acedemic & Professional. 1994, Chapter 2.

42 Liley, P.E.; Gambill, W.R. *Chemical Engineers' Handbook*, McGraw-Hill Book Co., New York, 1973. Section 3, pp 49.

43 Bertucco, A.; Piccinno, R., Soave, G. *Chem. Eng. Comm.* **1991**, *106*, 177-184.

44 Piringer, O. *Verpackungen für lebensmittel: Eignung, Wechselwirkungen, Sicherheit.* 1992. VCH, Weinheim, Germany. Chapter 6.1, pp 258.

Chapter 6

The Use of Model Substances for Migration Studies

Sue M. Jickells, Sue M. Johns, Katrina A. Mountfort, and Pilar Gonzalez Tuñon

Ministry of Agriculture, Fisheries and Food, CSL Food Science Laboratory, Colney Lane, Norwich NR4 7UQ, United Kingdom

Substances of differing volatility, polarity and molecular weight have been incorporated into plastics polymers and into cellulosic substrates and the migration of these model substances measured. These model substances act as surrogates for instrinsic migrants such as monomers, additives, contaminants etc. and permit considerable data to be obtained on factors influencing migration from a limited range of experiments. Examples are illustrated by the use of model substances to study transfer from board packaging to microwave heated foods; to develop a method of test for microwave susceptors and to evaluate plastics polymers as barriers to migration.

Migration studies can broadly be divided into two categories - surveillance/compliance type studies, carried out to check compliance with legislation or public health concerns and more investigative studies, such as those carried out to evaluate properties of newly developed polymers or the mechanisms influencing migration. For compliance-type studies, the migration of substances instrinsic to the food contact material is measured. Typically, the analysis will target one or two substances whose identity is known or, alternatively, materials may be analysed to identify constituents and contaminants.

Where the aim is to obtain mechanistic or performance information, following the migration of a number of substances can provide considerable information. Often these types of studies will be carried out using either one or two polymer types, varying the thickness, crystallinity etc. or they may involve a much wider range of polymers, again of different thickness, etc. For plastics or paper producers, life is simple because materials can be manufactured into which substances can be incorporated and migration subsequently measured. This is illustrated by the extensive migration studies carried out by Figge and co-workers (1-3) and researchers at Arthur D. Little (4-6), which were aided by the use of specially made polymers containing radiolabelled additives. However, even for producers, there can be difficulties incorporating some substances e.g. volatiles where extrusion/forming temperatures are high. For laboratories without access to primary production there are several choices - to arrange to have materials made, to incorporate substances oneself or to take pot luck on materials available from industry. The latter option permits a wide range of polymers to be obtained but does not guarantee which, if any, potential migrants may be present. It is expensive to have materials made and some producers are not keen to add substances not routinely used in production, particularly where they are producing commercial food contact materials on the same equipment.

Developing methods of analysis to detect and quantify migration is time consuming and expensive, particularly when measuring migration to complex matrices such as foods. Hence, although it is perfectly possible to use a range of food contact materials obtained from industry, there is no guarantee that potential migrants will be present and even less likelihood that the same migrants will be present in all materials at suitable concentrations. This leads to additional experimental work and added expense. The ideal situation is to be able to incorporate a variety of migrants of known identity into food contact materials at controllable levels and then to use these impregnated materials for migration studies. Sackett and co-workers used this technique, incorporating substances into susceptors and measuring their migration to foods (7). Begley and Hollifield used similar techniques for studying migration through susceptor films (8).

We report here techniques for introducing model substances into a variety of food contract materials and illustrate their use in migration studies.

Incorporation of model substances into cartonboard for microwave heating migration studies

There are a number of packaging situations where food is packed in paper and board for subsequent microwave reheating in the packaging by the consumer. If substances are present in the paper and board, either from manufacture or from printing, there is the potential for these substances to transfer to food. Studies were conducted to investigate whether substances transferred and, if so, was there a link with volatility and also whether certain packaging formats might reduce transfer.

Experimental. A schematic of the experimental procedure is shown in Figure 1. Microwave burgers (a hamburger-roll with a beef patty), packaged in cartonboard boxes were purchased frozen from retail stores. The boxes were disassembled and sprayed with a solution of model substances (see Table 1). Application was by airbrush spraying with an application rate of approximately 0.25 ml/dm². Crystal violet dye was added to the solution to enable visual evaluation of application during spraying. Blank samples were prepared by the same procedure using dye solution with no model substances. A cartonboard plaque (2 x 2 cm) was overlaid onto the packaging during spraying and subsequently analysed to determine the level of impregnation.

The boxes were reassembled and the food heated according to the manufacturer's on-pack instructions and then homogenised immediately after heating.

The model substances were extracted from the homogenised food using a combination of solvent extraction (acetonitrile/dichloromethane) followed by separation from the extracted fats using high performance-size exclusion chromatography (HP-SEC) (dichloromethane/cyclohexane mobile phase). Final analysis was by gas chromatography-mass spectrometry (GC-MS) in selected ion mode (SIM). The ions monitored for SIM are shown in Table I. Further details of extraction and analysis can be found in Johns *et al.* (9).

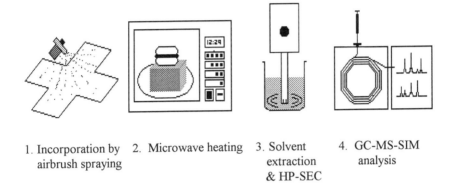

1. Incorporation by   2. Microwave heating   3. Solvent   4. GC-MS-SIM
    airbrush spraying   extraction   analysis
                                                   & HP-SEC

Figure 1. Procedure for microwave heating studies

Table I.   **Model substances, internal standards and ions monitored for GC-MS-**   **SIM analysis**

Substance[a]	MW[b]	B. Pt[c] (°C)	Function[d]	Ions for SIM (m/z)
Propyl benzoate	164.2	230	IS	105,123
1-Chlorodecane	176.7	223	MS	55, 91
Butyl benzoate	178.2	249	MS	105,123
Dimethyl phthalate	194.2	282	MS	77,163
d$_5$-Benzophenone	187.2	295[e]	IS	105,182
Benzophenone	182.2	305	MS	110,183
Benzyl butyl phthalate	312.4	370	MS	91,149
Diphenyl phthalate	318.3	255	IS	77,225

[a] Model substances prepared as a mixed solution in acetonitrile with each substance at 4 mg/ml
[b] MW=molecular weight; [c] B. Pt.=boiling point; [d] IS=internal standard, MS=model substance
[e] Calculated from value of 160°C/15 mm Hg. All other values at 760 mm Hg.

Three different retail packaging and heating configurations were examined (see Figure 2) to compare transfer for the different configurations and to see whether secondary packaging can act as a barrier to migration. Configuration one consisted of a burger heated directly on a susceptor in a cartonboard box with no paper bag. The second was a burger heated in a cartonboard box with no susceptor but with the burger wrapped in a paper bag. For the third configuration, the burger was heated in a paper bag on top of a cartonboard box. Where burgers were heated inside the box, the model substances were sprayed on the inside of the box. For the burger heated on top of the box, the substances were sprayed onto the outside of the box. Studies on cartonboard printed with UV-cure inks (9), showed that benzophenone, a photoinitiator used in UV-cure inks, was distributed throughout all layers of the board, indicating that low molecular weight substances applied to the outside during printing transfer readily throughout the board. It was therefore considered acceptable to spray the substances directly onto the food contact surface.

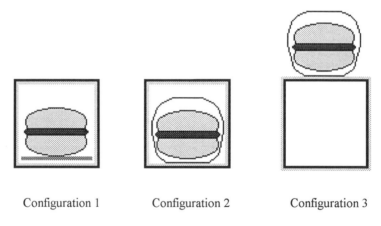

Configuration 1   Configuration 2   Configuration 3

Figure 2. Configuration of burgers for heating

Results and discussion. The transfer of model substances is shown in Figure 3. Values are presented as the percentage of the model substances incorporated into the board which transferred to food. The results are averages of triplicate heating of each food/packaging configuration. Agreement for replicate subsamples of foods was within $\pm22\%$. Transfer was highest to the burger heated in the box on a susceptor with no paper bag present. The presence of a paper bag reduced migration but there was still some transfer through the bag. Migration was lowest to the burger cooked on top of the box. This is probably to be expected as the volatilized model substances would be swept out of the microwave oven in the forced air flow and hence would have minimal contact time with the food. For burgers heated in the box, the more volatile substances transferred to the food to a greater extent, with 6-8% of 1-chlorodecane transferred compared to <1% benzyl butyl phthalate.

The model substances were selected to mimic the transfer of low molecular weight substances potentially occurring in cartonboard materials through production or printing. Benzophenone was chosen because it is used as a photoinitiator in UV-cure printing inks. Sebacate esters were selected because they can be used as plasticizers in printing inks. Phthalates can also be used as plasticizers in printing inks or can be present in paper and board materials as environmental contaminants. Although the model substances incorporated were not necessarily identical to substances in actual commercial use, they were, in general, chemically similar, covered a similar boiling point range and hence allowed conclusions to be drawn about the transfer of potential contaminants.

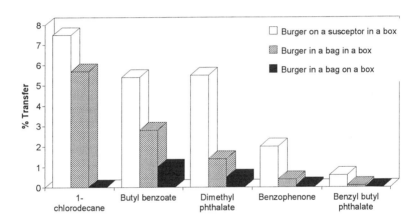

Figure 3. Transfer of model substances into burgers

Conclusions. The studies showed that for certain packaging and heating configurations, low molecular weight substances transfer readily to foods, with more volatile substances transferring more efficiently. The studies also showed that paper does not necessarily serve as a barrier to transfer. This knowledge on the relationship between volatility and transfer can be used together with screening-type analyses (solvent extraction and GC-MS analysis) to assess the potential for migration from retail packaging samples.

Development of a test for microwave susceptors

Microwave susceptors are used in the microwave heating of foods to promote browning and crisping. Currently, there is no recognised test for evaluating the safety-in-use of these materials, although several possibilities have been proposed (*10-15*). Tenax, a highly adsorbent powdered polymer of diphenylene oxide, has been adopted in Europe as a test medium for measuring migration from plastics materials and articles intended for use at high temperatures (*16, 17*) and has been used in other studies to trap migrants from microwave susceptors (*18*). We therefore explored the possibility of using Tenax as a test simulant for susceptors.

We have developed a test in association with the Fraunhofer Institute, Munich, Germany. The test has been validated by comparing migration into foods with migration into Tenax used as a simulant. In order to maximise the information for comparison, the migration of model substances was measured. Model substances were selected to cover a range of volatility and polarity (see Table II). Chlorinated hydrocarbons were used to avoid possible interference from non-chlorinated analogues generated during microwave heating (*18*). Butyl benzoate, benzophenone and benzylbutyl phthalate were selected as potentially being present from printing inks. Diethylene glycol dibenzoate was included as it has been shown to be used in laminating adhesives and to migrate from susceptors to foods (*19*).

Table II. Model substances used in susceptor studies

Substance	Incorporation technique	Boiling point (°C)	MW	Ions for SIM (m/z)
'volatiles'	Vapour phase equilibration			
Chlorobenzene		132	113	77, 112
3-Chloropropan-1-ol		161	95	58, 57, 76
1-Chlorononane		203	163	91, 92, 105
'non-volatiles'	Airbrush spraying			
Cyclohexylbenzene		240	160	160, 104
Butyl benzoate		249	178	105, 123
Benzophenone		305	182	182, 105
Benzylbutylphthalate		380	312	149
Diethylene glycol dibenzoate		410	314	105, 149
'internal standards'	Not applicable			
Fluorononane		168	146	97, 83, 85
1,9-Dichlorononane		260	197	91, 69

Experimental. Foodstuffs in packaging that incorporated microwave susceptors were purchased from retail stores. Foods were removed from the packages and stored at -20°C until required. The susceptors were discarded. Replacement susceptors, supplied by industry, were used for incorporation of model substances. The susceptors consisted of poly(ethylene terephthalate) (PET) film (vacuum sputtered with aluminium) laminated to cartonboard. Model substances were incorporated by airbrush spraying onto the cartonboard layer or by vapour-phase equilibration, depending upon the volatility of the substances (see Table II). Foods were placed on the impregnated susceptors and heated according to the manufacturer's instructions in a microwave oven with 700W nominal power. For the determination of the migration of non-volatile substances, the foods were homogenised after heating. Portions of homogenized food were then shaken with solvent and 1,9-dichlorononane internal standard, centrifuged and then a portion of the supernatant cleaned up by SEC. The fraction containing the substances of interest was collected and analysed by GC-MS-SIM. For the determination of the migration of volatile substances, foods were homogenised with chilled distilled water. Portions of the resultant slurry were transferred into headspace vials together with fluorononane as internal standard and then analysed by headspace GC-MS-SIM. The degree and homogeneity of incorporation of model substances into susceptors was assessed by taking portions of the impregnated susceptors, adding internal standards, extracting by shaking with solvent and then analysing the extracts by GC-MS-SIM. Further details of incorporation, extraction and analyses can be found in Mountfort *et al.* (*10*).

Migration into Tenax was determined by cutting a circle (0.6 dm^2 area) of susceptor impregnated with model substances, placing in a petri dish and covering with Tenax (2.4 g). The petri dish was covered with a glass plate and then heated for 10 min in a conventional oven at 180°C after which the Tenax was removed and extracted by shaking with diethyl ether (15 ml) and internal standard (0.5 mg). Calibration standards were prepared by spiking a solution of the model substances together with internal standards into diethyl ether extracts of blank Tenax. The ether extracts were analysed by GC-MS-SIM using a Hewlett-Packard 5890 Series II GC with a 5971A MSD operated in EI mode. Injections (1 µl) were made in splitless mode (splitless time 0.75 min) onto a 50 m x 0.25 mm i.d. fused silica capillary column with a 0.25 µm thick film of dimethylpolysiloxane. For non-volatiles the column was held at 80°C for 1 min after injection and then programmed to rise at 10°C/min to 300°C and held for 5 min. For volatiles the column was held at 40°C for 3 min after injection and then programmed to rise at 10°C/min to 150°C and then at 30°C/min to 250°C. The GC injector temperature was 250°C. Ions monitored are shown in Table II. The dwell time was 25 ms/ion.

Results and discussion. Migration results for pizzas, chips (french fries) and Tenax are shown in Tables III and IV. Values are the mean of heating 3 samples and with each then subsampled and analysed in triplicate. The repeatability for measurement of migration to food and Tenax was in the range ±10% (chips, Tenax) to ±17% (pizza). There is a complex relationship between the percentage migration and the boiling point of the model substances. For both chip and pizza susceptors, migration to food and to Tenax tends to be very low for the low molecular weight-high volatility substances. Migration then rises steeply to cyclohexylbenzene and then declines steadily as the boiling point increases further. Figure 4 illustrates this linear trend at the higher molecular weight end of the range.

Table III. Transfer of model substances to chips and to Tenax

Model substance	Migration to Tenax (%)	Migration to chips (%)	Ratio Tenax/food
3-Chloropropan-1-ol	nq	nq	-
Chlorobenzene	7	5	1
Chlorononane	8	2	4
Cyclohexyl benzene	62	15	4
Butyl benzoate	60	11	5
Benzophenone	46	5	9
Butylbenzylphthalate	26	2	13
Diethylene glycol dibenzoate	11	1	11

nq - not quantifiable

Table IV. Transfer of model substances to pizza and to Tenax

Model substance	Migration to Tenax (%)	Migration to pizza (%)	Ratio Tenax/food
3-Chloropropan-1-ol	3	3	1
Chlorobenzene	7	6	1
Chlorononane	4	2	2
Cyclohexyl benzene	52	4	13
Butyl benzoate	50	4	12
Benzophenone	36	3	12
Benzylbutylphthalate	17	2	9
Diethylene glycol dibenzoate	nq	1	-

nq - not quantifiable

Migration of volatiles into Tenax was typically less than 10%. Previous work had found over 50% migration (*10*). The earlier work had used microwave heating whereas the present work used a laboratory oven - conditions more appropriate for a standard test since microwave heating is more difficult to make uniform between laboratories. The test time and temperature was selected as a result of visually comparing browning of the board layer of susceptors heated with a food load in a microwave oven, with the extent of browning of the same materials heated for varying times and temperatures in a laboratory oven with no food load (*10, 20*). For chips and pizza susceptors, heating in a laboratory oven for 10 min at 180°C gave a similar degree of browning to that produced by the microwave heating of these foods and hence this time and temperature was adopted for testing. The use of Tenax allied with conventional oven heating for testing susceptors has also been used by Booker and Friese (*18*), although they measured temperatures using fiber optic probes.

Experiments with thermocouples showed that the Tenax heated rather slowly on being placed in the hot oven (180°C), reaching between 144 to 155°C after 10 min. The susceptor, by contrast, which was in intimate contact with the heated petri dish, heated rapidly. Migration to Tenax (and to foods) will be a balance of the rate of volatilisation from the susceptor, the efficiency of trapping by the Tenax or food and finally the rate of subsequent desorption from the hot Tenax or food. Presumably the more volatile substances were volatilized from both the susceptor and the Tenax (and the food) whereas the less volatile substances were volatilized from the hot susceptor but were not been as readily desorbed from the Tenax, which was at a lower temperature. A reduction factor was calculated for each substance i.e. the ratio of migration into Tenax versus the migration into food. Reduction factors ranged from 1 to 13 for both chip and pizza susceptors (Table III and IV respectively).

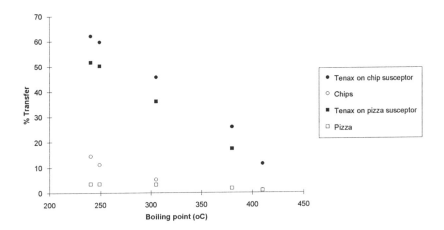

Figure 4. Migration of non-volatile substances into foods and Tenax simulant

Conclusions. For assessing the safety-in-use of susceptor materials, the ideal scenario would be to measure migration to foods. However, this is difficult analytically and hence it is commonplace to measure migration into simulants. The ideal simulant is one which gives the same migration as foodstuffs. Measuring the transfer of a number of different model substances from susceptors has shown that the use of Tenax as a simulant will tend to overestimate migration, particularly for less volatile migrants. Overestimation is acceptable in that it does provide a safety margin but the margin should not be so large as to be unrealistic. A correction for overestimation can be made by applying reduction factors to the results of migration testing. For substances of boiling point greater than 180°C, a reduction factor of 10 or 15 would appear appropriate. For more volatile substances, a reduction factor does not seem appropriate. As a result of the work, a method for the migration testing of microwave susceptors has been established and has been submitted to the Comitté Européen de Normalisation (CEN) for adoption as a standard method.

Evaluation of polymers as functional barriers to migration

A functional barrier can be defined (*21*) as 'any integral layer, which under normal and foreseeable conditions of use, reduces all possible material transfer (permeation and migration) from any layer beyond the barrier into food to a toxicologically and organoleptically insignificant and technologically unavoidable level'. If a packaging material acts as a functional barrier this can be used in a number of ways. For example, there is considerable debate about the safety of recycling post-consumer plastics packaging if it will be used to package food. Recycling becomes a safer option when a functional barrier is used between the recycled material and the packaged food because, by definition, the barrier reduces transfer of substances to a safe level assuming, of course, that the barrier layer itself does not contaminate the food. Functional barriers can be used not only to prevent ingress of substances to food but can also be used to prevent loss of substances from foods. This can be important for flavour compounds whose loss from foods can lead to organoleptic changes in a product and ultimately to customer dissatisfaction.

We have been carrying out studies to evaluate polymers as functional barriers to migration. We have used model substances to investigate the influence of polymer type, thickness, temperature and time.

Experimental. Benzophenone, as a model substance, was incorporated into a polyethylene powder (low density, 300 μm particle size, density 0.92) which was then placed in contact with one side of the test film. A disc (7.6 cm diameter) of low density polyethylene film (LDPE) (density 0.92, 160 μm thick) was placed in intimate contact with the other side of the test film to act as a recipient for the substances. The migration studies were carried out in a specially designed migration cell illustrated in Figure 5, with a contact area of 0.25 dm^2. After exposure under defined conditions of time and temperature the donor powder, recipient film and test material were separated and then a portion of the recipient and test films each shaken with dichloromethane in the presence of internal standard (1-chlorodecane). The dichloromethane extracts were analysed by GC with flame ionisation detection (GC-FID) in split or splitless mode, depending on the level of migration. The GC-FID was a Carlo Erba HRGC 5300 series fitted with a 25 m x 0.25 mm i.d. fused silica capillary column with a 0.25 μm thick phase of dimethylpolysiloxane. Injections (1 μl) were made in split (split ratio 20:1) or splitless mode (splitless time 45 s). For split injections, the column was held at 140°C for 2 min after injection then programmed to rise at 40°C/min to 290°C and held at this temperature for 5 min. For splitless injections, the column was held at 40°C during injection and for two minutes thereafter then heated at 40°C/min to 290°C and held for 5 min. Helium was used as carrier gas with a column head pressure of 50 Kpa. The GC injector and detector temperatures were 290°C and 300°C respectively.

The benzophenone was incorporated into the PE powder by spiking a solution of the substance in diethyl ether (5 ml of a 0.03 mg/ml solution) into the power (100g) and shaking overnight on a roller-shaker. The level and homogeneity of incorporation was determined by extracting sub-samples (0.5 g) of the powder with dichloromethane (2 ml), in the presence of internal standard (1-chlorodecane). The extracts were analysed GC-FID in split mode.

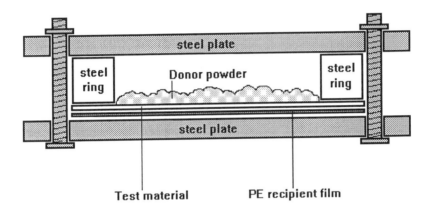

Figure 5. Migration cell used for functional barrier studies

Results and discussion. Benzophenone was incorporated into PE powder at 1.38 ± 0.036 mg/g (n=6). The target level for incorporation had been 1.5 mg/g and hence it is concluded that the method allows homogeneous incorporation at selected levels.
Migration of benzophenone through the test material was measured as uptake by the PE film recipient and expressed as a percentage of the substance present in the PE donor powder at t_0. Migration through the polymer materials tested at 40°C is shown in Figure 6.

Migration through a polymer layer can be considered to consist of 3 main phases: the initial 'lag phase' (22) when the migrant is diffusing through the barrier but has not permeated through the whole barrier; the diffusion phase, when the migrant is transferring from the barrier but migration is not concentration limited and is directly proportional to the square root of time ($t^{1/2}$); and the plateau phase, when migration becomes concentration limited. The data show that there was transfer through all of the polyolefin materials evaluated. Transfer through the thinner of the PE films tested (160 μm, density 0.92 g/cm^3, approximately 50% crystallinity) was rapid, with the lag phase for benzophenone between 0.25 and 0.75 h. Similar experiments carried out at 5°C showed a lag phase of between 18 and 24 h. These results indicate that a thin film of LDPE is a poor barrier and that it does not afford protection against migration of moderately volatile substances such as benzophenone, even at refrigerator temperatures. The lag phase was considerably longer for the thicker (1 mm) PE, with a lag phase of 14 h for LDPE (specifications as for the thinner PE film) and 79 h for high density PE (HDPE) (density 0.95 g/cm^3, 70-80% crystallinity).   PET was found to be a much better barrier to transfer of benzophenone than the polyolefin films tested.   There was no transfer of benzophenone through the 12 μm thick Melinex PET film even after 2 weeks at 40°C. There was no measurable migration through the PET film after 1 week at 70°C but after a further week at 70°C, 0.1% of the available benzophenone had migrated. Replacing the PE donor powder with PET powder impregnated with benzophenone did not result in higher transfer, indicating that the partition from the powder into the PET film was not a rate limiting step.

The experimental procedure described can be used to evaluate polymers as functional barriers.   Although the studies reported used benzophenone as a model substance, other substances could be used, alone or in admixture.   The use of the polymer powder allows a variety of substances to be tested against the barrier at selected concentrations.   The test cell can be readily adapted to replace the film recipient with foods or foods simulants so that the effect of these parameters on migration can be studied.

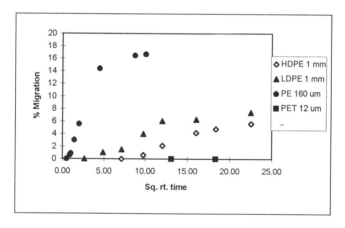

Figure 6.   **Migration of benzophenone through polymer films at 40°C**

Conclusions

Model substances can be incorporated into polymers and cellulosic substrates and used for migration studies.   The techniques for incorporation described permit substances with differing volatility, molecular weight and polarity to be incorporated so that data obtained are applicable to a wide range of potential migrants.   Our studies have typically used low molecular weight hydrocarbons and esters as these types of substances are used in printing inks and hence may be present in printed food contact materials. However, the techniques can be applied readily to other substances, although it may be necessary to use analytical methods other than GC and GC-MS depending upon the substances incorporated.

Acknowledgements

The authors wish to acknowledge the British Ministry of Agriculture, Fisheries and Foods for provision of funding for the work on microwave heating and microwave susceptors and the Leonardo programme of the European Union for the fellowship for Pilar Gonzalez Tuñon.

Literature Cited

(*1*)   Figge, K. and Freytag, W. *Food Add. Contam.* **1984**, *1*, 337-347.

(*2*)   Bieber, W.-D.; Figge, K. and Koch, J. *Food Add. Contam.* **1985**, *2*, 113-124.

(*3*)   Bieber, W.-D.; Freytag, W.; Figge, K. and vom Bruck, C.G. *Food Chem. Toxic.* **1984**, *9*, 737-742.

(*4*)   Schwope, A.D.; Till, D.E.; Ehntholt, D.J.; Sidman, K.R. and Whelan, R.H. *Food Chem. Toxic.* **1987**, *4*, 317-326.

(*5*)   Schwope, A.D.; Till, D.E.; Ehntholt, D.J.; Sidman, K.R.; Whelan, R.H.; Schwartz, P.S. and Reid, R.C. *Deut. Lebens.-Rund.* **1986**, *82*, 277-282.

(*6*)   Goydan, R.; Schwope, A.D.; Reid, R.C. and Cramer, G. *Food Add. Contam.* **1990**, *7*, 323-337.

(*7*)   Sackett, P.H.; Wendt, D.J.; Graff, J.T. and Pesheck, P.S. *Tappi J.*, November **1991**, 175-181.

(*8*)   Begley, T.H. and Hollified, H.C. In *Food and Packaging Interactions II*; Risch, S.J. and Hotchkiss, J.H., Eds.; ACS Symposium Series 473, ACS, Washington DC, 1991, pp 53-66.

(*9*)   Johns, S.M.; Gramshaw, J.W.; Castle, L. and Jickells, S.M. *Deut. Lebens.-Rund.* **1995**, *91*, 69-73.

(*10*)   Mountfort, K.; Kelly, J.; Jickells, S.M. and Castle, L. *J. Food Prot.* **1995**, *59*, 534-540.

(*11*)   Hollifield, H.C. In *Food and Packaging Interactions II*; Risch, S.J. and
         Hotchkiss, J.H., Eds.; ACS Symposium Series 473, ACS, Washington DC,
         1991, pp 22-36.

(12)   ASTM. 1990. American Standard F-1308-90. Standard test method for
       quantitating volatile extractables in microwave susceptors used for food
       packaging. American Society for Testing Materials, Philadelphia, PA, USA.

(*13*)   ASTM. 1993. American Standard F-1349. Standard test method for
         nonvolatile ultraviolet (UV) absorbing extractables from microwave
         susceptors. American Society for Testing Materials, Philadelphia, PA, USA.

(*14*)   ASTM. 1993. Draft American Standard. Test method for quantitating non-
         UV absorbing extractables from microwave susceptors utilizing solvents as
         food simulants. American Society for Testing Materials, Philadelphia, PA,
         USA.

(15)   Rose, W.P. In *Food and Packaging Interactions II*; Risch, S.J. and Hotchkiss,
       J.H., Eds.; ACS Symposium Series 473, ACS, Washington DC, 1991, pp 67-
       78.

(16)   Piringer, O.; Wolff, E. and Pfaff, K. *Food Add. Contam.* **1993**, *10*, 621-630.

(17)   Commission Directive 97/48/EC of 29 July 1997. *Off. J. Eur. Comm.* **1997**. *L
       222*, 10-15.

(18)   Booker, J.L. and Friese, M.A. *Food Technol.* **1989**, *43*, 110-117.

(19)   Begley, T.H.; Biles, J.E. and Hollifield, H.C. *J. Agric. Food Chem.* **1991**, *39*,
       1944-1945.

(20)   Wolff, E. 1994. Final report to European Commission DGIII-C-I for contact
       ETD/92/B5-300/M1/66.

(21)   Council of Europe document CD-P-SP(93) 25 (18 October 1993).

(22)   Crank, J. and Park, G.S. *Diffusion in Polymers*; Academic Press, London,
       1968.

Chapter 7

Fatty Food Simulants: Solvents to Mimic the Behavior of Fats in Contact with Packaging Plastics

A. E. Feigenbaum[1], A. M. Riquet[2], and D. Scholler[1,3]

[1]Institut National de la Recherche Agronomique SQuAIE,
Moulin de la Housse, 51687 Reims, France
[2]Institut National de la Recherche Agronomique,
Domaine de Vilvert, 78352 Jouy en Josas, France
[3]UFR Sciences, 51687 Reims Cedex, France

Fatty food simulants play an essential role in safety assessment of food packaging materials. Official simulants (olive oil, corn oil) are a source of difficulties for migration determinations. Regulations accept the use of solvents to replace fats in migration tests. However, interactions of ethanol and isooctane -two widely used solvents with plastics- largely differ from those of fats. Criteria for selection of simulants are discussed, on the basis of thermodynamic and kinetic parameters, using paramagnetic and optical probes in polymers. We show here how the aggressiveness of solvent mixtures can be tailored to mimic fats in contact with widely used packaging plastics.

Work on the understanding of food and packaging interactions at INRA started in 1985. Its aim was to design materials with reduced migration and to provide a scientific basis for developing EU regulations on plastics in contact with food. Part of this work was developed in the framework of the European research programme AIR 941025, coordinated by INRA. Test media made from mixtures of solvents, whose principle is explained here, were authorized by the 97/48/EC directive (1), as a consequence of this work.

Since foodstuffs can give rise to difficulties when used for migration testing, most regulations allow the use of food simulants. However, fatty food simulants are triglycerides (olive oil, corn oil, medium chain triglycerides), and migrant analysis in such media is often difficult. Hence, the need for alternative volatile test media which have been proprosed in literature and accepted in scientific regulations including heptane, isooctane, ether, ethanol and isopropanol. The very number of proposals still proposed in recent years (2) suggests that none of them is really satisfactory.

In 1997, the Commission laid down the basic rules for migration testing (1), thus updating the 82/711/EEC directive. The official fatty food simulant is olive oil. When its use is not possible for practical reasons, substitute tests can be carried out, using isooctane or ethanol as test media. For routine testing (alternative tests), any solvent or mixture of solvents can be used, as long as migration is at least as severe as that measured in olive oil. The purpose of the present chapter is to explain how these mixtures can be defined. In order to do this on a scientific basis, we will first review food and packaging interactions.

1. LIQUID AND PACKAGING INTERACTIONS

1.1 Polymer swelling by the food simulant (or the extractant)

Figure 1 shows two mass transfer phenomena occurring during contact of a polypropylene random copolymer tray (2 mm) with isooctane: sorption of the solvent by the polymer (figure 1a) and migration of aromatic antioxidants from packaging to food simulant (figure 1b). Sorption is monitored by the percentage of weight uptake of the material, and migration by the absorbance at 275 nm (3, 4). Plots against the square root contact time display a so called " non-Fickian " behaviour. If these phenomena were Fickian, the part of the curve before the steady state would be a straight line, whose slope would yield the diffusion coefficient. The upwards curvature of the experimental sorption curve indicates a constant increase of the diffusion coefficient: penetration of isooctane molecules facilitate further penetration by plasticization of the polymeric matrix, until the plateau is reached (64 h). Plasticization also influences migration: the diffusion coefficient is initially very small, and then increases until the material reaches equilibrium with isooctane. Clearly, solvent penetration is the driving force of mass transfer into the test media and is the key to its aggressiveness. The latter could be quantified by its solubility in the polymer, given by the solvent uptake at equilibrium (table 1)

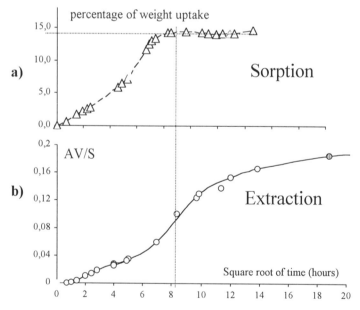

Figure 1:a) solvent sorption (weight of the material m_0 and m_t at times 0 and t respectively) (upper curve) and b) migration of aromatic antioxidants (absorbance A, volume of solvent V, area of film S) (lower curve) from a PP random copolymer 2 mm tray

Solvent (mol. weight, density)	Weight uptake (g/100 g polymer)	Volume absorbed (mL/g polymer)	Mmoles absorbed (mmoles/g polymer)
Dichloromethane (M=84, d=1.36)	23	0.169	2.7
Diethyl ether (M=74, d=0.69)	13	0.18	1.75
Isooctane (M=114, d=0.71)	14	0.202	1.22

Table 1: aggressiveness of solvents to PP random copolymer (5% ethylene), based on various parameters

1.2 Migrant accessibility: influence of polymer heterogeneity.

The importance of accessibility on migration was already pointed out by Adcock (5). Convex sorption isotherms (Flory-Huggins, dual mode) reflect the heterogeneity of polymeric matrixes (6). The most famous example is vinyl chloride (VCM) whose partition coefficient between food and packaging ($K_{F/P}$) decreases as the VCM concentration decreases in polymer. Some VCM molecules are loosely bound, or are close to the surface and they migrate first. Others are trapped in the structure and cannot migrate. Sites with intermediate situations exist, which explains the gradual change of $K_{F/P}$ (7).

Another example has been shown recently with bisphenol A diglycidyl ether (BADGE), a starting substance for vinylic organosols used as inner can coatings (8). These coatings are made of two interpenetrated polymers: PVC and an epoxyphenolic (EP) phase. BADGE is used in large excess, which lowers the glass transition temperature of the PVC phase. Using refluxing methanol, up to 4 mg BADGE/dm^2 could be extracted. At 40°C, in migration experiments [isooctane + tert-butyl acetate (0-40%)] and even in extraction tests [isooctane + tert-butyl acetate (40-100%)], not more than 2 mg BADGE/dm^2 was available. This could be ascertained by using paramagnetic probes, incorporated into the coating before curing. The probes were located both in the EP and in the PVC phases, and the corresponding ESR (Electron Spin Resonance) signals were well differentiated. After treatment with aggressive solvents ($\square$ 40 % tert-butyl acetate in isooctane) at 40°C, the probes of the EP phase were still present, while those of the PVC phase were quantitatively extracted. This selective extraction from the PVC phase is also likely to occur for BADGE, since extraction at 40°C resulted in an increase of the glass transition temperature of PVC.

The strong temperature dependence observed for vinylic organosols (40°C compared to boiling methanol) suggest that when accelerated tests are needed, one should use caution when tampering with temperature, which may modify the structure of the polymer. Furthermore, accessibility effects are likely to be highly dependent on sterical hindrance, and should play a major role with bulky triglycerides, like HB 307 or olive oil.

1.3 Simulant selectivity: solvent and packaging partition.

In section 1.2 we have shown the importance of heterogeneity of polymeric matrices. In this section we will consider the influence of the interaction of a migrant with the solvent or the food simulant. It can be assumed that a migrant partitions at steady state between the food simulant and the polymer. Since the volume of food is much larger than that of plastic, the trend is that a large fraction of migrant ($M_{F,\infty}/M_{P,0}$) will be transferred to the food. This fraction can be calculated as a function of the partition coefficient $K_{F/P}$ and of the volumes of food and of plastic. In standard EU test conditions ($1 dm^3$ food in contact with 6 dm^2 packaging), $M_{F,\infty}$ depends on $K_{F,P}$ and on the thickness of the package L_p (figure 2). It appears that even for thick materials, as soon as the partition coefficient $K_{F/P}$ exceeds 0.5, more than 80% of the migrant is transferred to food at equilibrium (9).

Percentage of migration/100

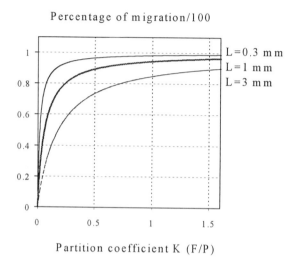

Partition coefficient K (F/P)

Figure 2: influence of the partition coefficient $K_{F/P}$ on the migration at equilibrium (assuming standard EU conditions: 6 dm^2 in contact with 1 dm^3 food) (19)

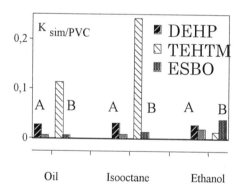

Figure 3: Partition coefficients of three plasticizers of two PVC samples A (DEHP and ESBO) and B (TEHTM + ESBO) and three simulants (olive oil, isooctane and ethanol) DEHP = bis(2-ethylhexyl) phthalate, TEHTM = tris(2-ethylhexyl) trimellitate, ESBO = epoxidised soybean oil.

A food simulant must mimic the behaviour of fats for all the migrants of the given material. Nevertheless, all migration studies reported in literature have dealt with a single migrant. This was the case in all studies with radiolabelled additives (10, 11) or with global migration (12) (often, global migration is nothing else than the specific migration of a major migrant: oligomers for polyolefins, plasticisers for plasticised PVC, mineral oil for polystyrene). One cannot assume the same behavior with all the migrants in a mixture, as shown in the example reported in figure 3, which displays migration of three substances from plasticised PVC: bis(2-ethylhexyl) phthalate (DEHP), the lipophilic tris(2-ethylhexyl) trimellitate (TEHTM) and epoxidized soybean oil (ESBO). Two PVC samples (A) and (B), one containing DEHP and ESBO (A), the other containing TEHTM and ESBO (B), were submitted to mass transfer into olive oil, ethanol and isooctane.

At equilibrium, DEHP partitions in the same manner into all three test media. However taking into account only this single migrant is misleading. If one also considers the behaviour of the two other migrants, it appears that ethanol is too polar, it has too much affinity with ESBO and not enough with TEHTM, and it does not reflect the behaviour of olive oil (figure 3). Isooctane is much closer to olive oil.

This demonstrates the need to take into account a range of different affinities which can be done by considering solubility parameters. Figure 4 displays a chart of solubility coefficients, adapted from Van Krevelen (13, 14). Olive oil is neither close to ethanol nor to isooctane.

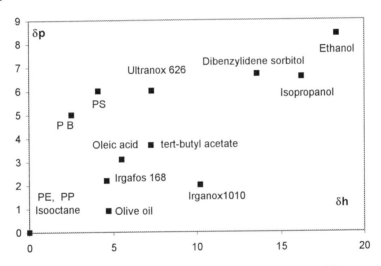

Figure 4: Solubility parameters of polymers (see glossary for abbreviations), of additives and of food simulants. (based on ref. 13 and 14)

Hence, the idea that a simulant whose solubility parameter is close to that of olive oil could be made by mixing a volatile ester (olive oil interacts through its ester groups) and isooctane. Such a test medium would be suitable for polar polymers (PVC, PET, polycarbonate). However, isooctane has an exaggerated swelling effect on apolar polymers, and in this case the non aggressive component of the simulant mixtures should be an alcohol: ethanol or isopropanol. The latter is an interesting medium to be studied in future studies, since it is much less reactive than ethanol with additives (12). However, since ethanol is recommended by most regulations, we used this medium in the present work.

2. THE TOOLS USED

Diffusion coefficients of solvents in polymers were determined by monitoring either the weight uptake of the material or the $v_{C=0}$ absorbance of the esters by infra-red (1740 cm^{-1}).

Diffusion coefficients of olive oil in polyolefins were obtained from the concentration profiles in the material, as measured on transverse sections, using FTIR microscopy ($v_{C=0}$=1740 cm^{-1}).

Changes in free volume in the polymer were observed using either methyl red as an optical probe, or aminoxyls as paramagnetic probes.

The spectrum of methyl red in dichloromethane is driven by an intramolecular hydrogen bond which compensates steric repulsion with ortho H atoms, and which ensures planar conformation with a strong extinction coefficient (λ_{max} 488 nm; $\varepsilon = 44000$ mol^{-1} x L x cm^{-1}). In esters like ethyl acetate, the hydrogen bond takes place with a solvent molecule, allowing a release of steric hindrance in a twisted conformation. This justifies a bathocromic shift, with a lower extinction coefficient ($\lambda_{max} = 478$ nm; $\varepsilon = 37000$ mol^{-1} x L x cm^{-1}).

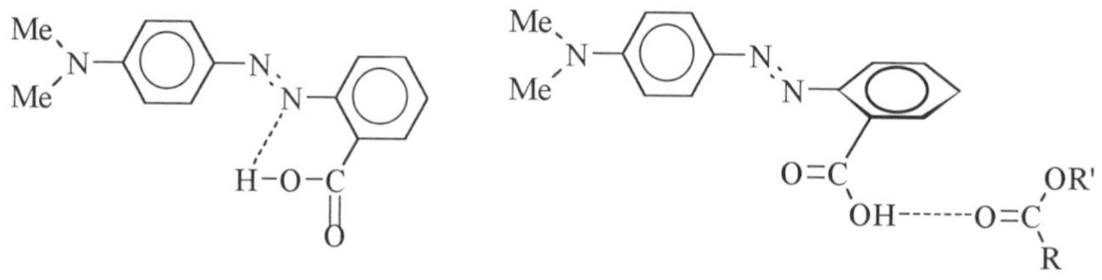

Figure 5: Methyl red conformations: left: in dichloromethane, and in rigid PVC ($\lambda_{max} = 488$ nm); right : in PVC swollen by fatty acid esters RCOOR' ($\lambda_{max} = 478$ nm)

In polymers swollen by fatty esters, the spectrum of methyl red depends on the extent of swelling. With long chain esters, swelling is very slight, and the molecules lack space to find the right arrangement to enable hydrogen bonding. With short chain esters, there is a strong swelling, which favors hydrogen bonding and explains the bathocromic shift and the drop of the extinction coefficient (15) (figure 5). What is remarkable with this probe is that its sensitivity (parallel to ε) is highest when there is a very slight swelling, which makes it an excellent tool for the exploration of food and packaging interactions.

Similar free volume effects are observed with paramagnetic probes. When the probe is immobilised in the polymeric network, the spectrum is broad and disymetric (lines H$_{-1}$, H$_0$, H$_{+1}$,) (figure 6a). When the aminoxyl probe can move freely around the direction of the magnetic field of the instrument, a narrow spectrum made of three sharp lines (lines h$_{-1}$, h$_0$, h$_{+1}$,) is observed (figure 6b). The broad disymetric signal and the narrow sharp triplet are usually observed in polymers respectively below and above their glass transition temperatures. When a material is in contact with fatty food simulants, a composite spectrum is usually observed (figure 6c), reflecting a heterogeneous distribution of the probes in the material: some probes, close to simulant molecules, are surrounded by a larger free volume, resulting in the narrow sharp triplet even well below the glass transition temperature of the polymer. This effect is linked to local plasticization of the polymer by the simulant (16, 17). The ratio of line intensities h$_{-1}$/H$_{-1}$ represents an evaluation of the plasticization of polymers by solvents. The structures of the probes studied are given in figure 7.

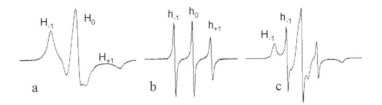

Figure 6: ESR spectra of aminoxyl (NO°) probes (a) immobilized in a polymeric matrix (e.g. below glass transition), (b): fast moving (e.g. above Tg); (c): in a polymer penetrated by a simulant

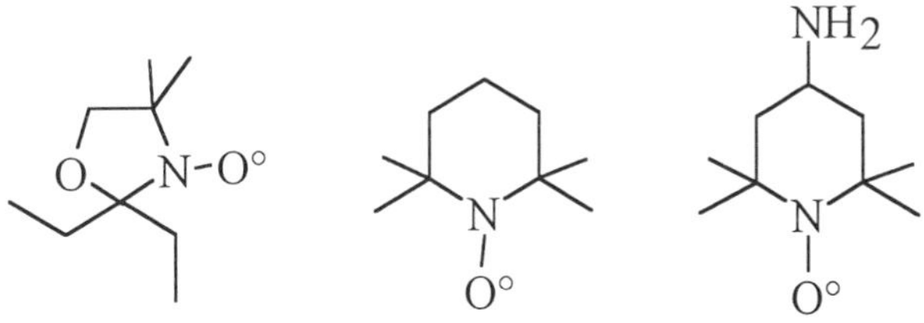

1   **2**   **3**

Figure 7: paramagnetic probes used for the study :
1=2-DOXYLpentane; 2 =2,2,6,6-tetramethylpiperidin-1-oxy;
3= 4-amino-2,2,6,6-tetramethylpiperidin-1-oxy. 3 is the more polar probe

3. DESIGN OF MIXTURES OF SOLVENTS

3.1 Isooctane-tert-butyl acetate mixtures for rigid PVC

Probes **1-3** (figure 7) were incorporated separately in rigid PVC. PVC samples (2 cm x 1 cm) were immersed in simulants made from isooctane and tert-butyl acetate, in different proportions. Several parameters were monitored together as a function of the percentage of ester:
- the ratio h_{-1}/H_{-1} (figure 8a),
- the weight uptake (figure 8b),
- the migration of the probes (figure 8c) after 24h.

All these parameters change in the same way: with pure isooctane, no interactions can be detected. With 5-10% tert-butyl acetate, both migration and weight uptake slightly increase. Above 20%, there is a sudden break in the curves for all three parameters, indicating a critical composition of the simulant which plasticizes the polymer. With pure tert-butyl acetate, the PVC is completely swollen and almost dissolved.

Medium chain triglycerides (synthetic triglycerides mixture like HB 307) also slightly plasticize PVC, and give rise to migration. It is shown on figures 8b and 8c that they behave like 10% tert-butyl acetate isooctane mixtures.

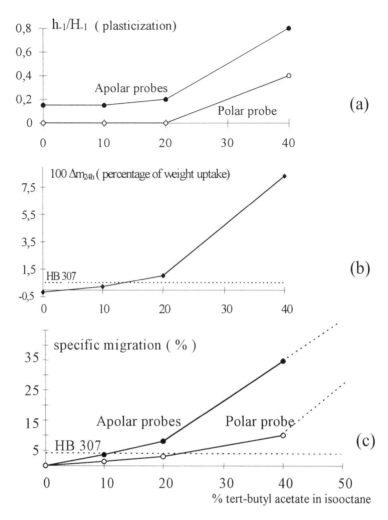

Figure 8: evaluation of the aggressiveness of mixtures of tert-butyl acetate and isooctane (a): by ESR; (b): by weight uptake; (c): through migration of paramagnetic probes

Probe **3** showed specific behavior, to the extent that it migrated more slowly than other probes, even if they had a higher molecular weight. This behavior was shown to be linked to hydrogen bonding with other additives (like ESBO) or with solvent molecules. The diffusing species was not **3**, but a complex, associating **3** with a hydrogen acceptor molecule of the solvent, like tert-butyl acetate.

3.2 A colorimetric test for the selection of a simulant for rigid PVC.

When solutions of ethyl acetate in isooctane containing methyl red (0.015 g x l⁻¹) were left in contact with rigid PVC, a similar trend to that observed in figures 8b-c was observed. No discoloration was observed with pure isooctane. However when the concentration of tert-butyl acetate exceeded 5 %, the sample was discolored. It could be shown that the penetration of the dye in PVC was equivalent to that of the simulant, as they had very close diffusion coefficients. The dye alone did not penetrate without the simulant. The driving force for its penetration was that of the simulant. This enables using a simple test to detect the critical concentration which plasticizes the material. Around this concentration, it is possible to tailor the aggressiveness of the mixture: depending on the percentage of ester, one can obtain a medium which behaves like oil, or like a mixture of medium chain triglycerides. The colorimetric method is a good one in this region due to its good sensitivity in the neighborhood of the critical concentration, where the interaction is still weak.

3.3 Ethanol-tert-butyl acetate mixtures for polyolefins.

Isooctane interacts too strongly with polyolefins and must be replaced by ethanol or isopropanol. Ethanol should be avoid with reactive migrants (epoxides give addition, esters give transesterification or degradation (12). Since it is recommended in most regulations, we have used ethanol as solvent in the mixtures.

The penetration of ester-based simulants could be monitored here by Fourier Transform Infrared Spectroscopy (FTIR). When possible, this method is simpler than the use of paramagnetic probes, which requires special manufacturing. The penetration of tert-butyl acetate - ethanol mixtures was monitored by transversal FTIR spectroscopy, for different compositions. In contrast to PVC (figure 8), the penetration increased linearly with the ester percentage up to 40%. The material being rubbery, there is no breaking point corresponding to a critical plasticizing concentration.

As suggested in section 1.1, the aggressiveness of a medium towards a polymer could be defined by its solubility in the material. In order to determine the solubility of olive oil in a 2 mm PP tray , a direct infrared method was not practical: a complete equilibration of the material with olive oil would require 5 years. The use of a film instead of a tray may require a material with a different polymer grade. In order to use the same material, we analysed cross sections of the tray by FTIR microscopy, and determined, at different times the concentration of olive oil at different depths from the surface. The correspondence of experimental data with calculated curves gave the kinetic and thermodynamic parameters (18): $D = 7.5 \times 10^{-11}$ cm^2 . s^{-1}

C_{olive}^{PP} = 18.9 μmol x cm^{-3} (using an average molecular weight 960 g x mol^{-1}). This value is equivalent to 5 % tert-butyl acetate in ethanol which simulates the behavior of olive oil.

In conclusion, it is possible to define the aggressiveness of solvents and of fats towards polymers, and to define equivalencies, based on solubility of these media in the plastics, as shown in table 2.

	Rigid PVC tert-butyl acetate in Isooctane	Polypropylene[a] tert-butyl acetate in ethanol	Polyethylene[b] tert-butyl acetate in ethanol
Olive oil			4 %
HB 307	10 %		
Miglyol 812	15 %	10 %	9 %

Table 2: Mixtures of solvents having the same solubility as fatty media; a: random copolymer; b: low density ; HB 307 : synthetic triglycerides mixture (10% C10-50% C12 ; Miglyol 812 : synthetic triglycerides mixture (55% C8-45% C10)

ACKNOWLEDGEMENT

This paper reviews work which has been conducted partly in the frame of the European research programme AIR 941025 (1994-1997). It has also links with other programmes at INRA: with Pechiney (Ph.D. of Sylvie Cottier), with DGER, ENSIA (Ph.D. of Patricia Métois), DGAL and with Europol'Agro. The authors thank these institutions and companies.

GLOSSARY

Polymers: PB, PE, PET, PP, PS, PVC, EP = polybutadiene, polyethylene, poly(ethylene terephthalate), polypropylene, polystyrene, polyvinyl chloride and epoxyphenolic respectively.
Additives cited: Irgafos 168 = tris[2,4-ditert-butylphenyl] phosphite;
                      Irganox 1010 = pentaerythritoltetrakis[3-(2,6ditert-butylphenyl)
                      propionate
                      DBS= Dimethylbenzylidene sorbitol
      INRA = Institut National de la Recherche Agronomique
      SquAlE = Equipe "Sécurité et Qualité des Emballages Alimentaires"

REFERENCES

(1)   EUROPEAN COMMUNITY 1997: 97/48/EC Council directive: modification of 82/711EC directive establishing rules for migration testing; Official Journal of European Communities, L. **222**, 12.8.1997, Brussels, Belgium

(2)   ROSSI L. 1995/96: History of the fat test and its future at the European level, Journal of Polymer Engineering **15**, 17 -32

(3)   FEIGENBAUM A.E., BOUQUANT J., HAMDANI M., METOIS P., RIQUET A.M., SCHOLLER D.,1997: Quick methods to control compliance of plastic materials with food packaging regulations, Food Additives and Contaminants, **14**, 571-582.

(4)   METOIS P., SCHOLLER D., BOUQUANT J., FEIGENBAUM A., 1998: Alternative test methods to control the compliance of polyolefin food packaging materials with the European Union regulation: the case of aromatic antioxidants and of bis(ethanolamine) antistatics, based on ^{1}H-NMR and UV-visible spectroscopy, Food Additives and Contaminants, **15**, 100-111

(5)   ADCOCK L.H., HOPE W.G., SULLIVAN D.A., WARNER A.M. 1984. The migration of non-volatile compounds from plastics. Part3. Further experiments with model systems and development of the descriptive and pictorial concept of migration, Plastics and Rubber Processing and Applications, **4**, 53-62.

(6)   De V. NAYLOR T., 1989: Permeation properties; in Comprehensive Polymer Science, **2**, C. Booth and C Price ed, Pergamon, Oxford.

(7)   MILTZ J., 1992: Food Packaging, Handbook of Food Engineering, edited by D. Heldman and D. Lund, Marcel Dekker, New-York, p. 667-718

(8)   COTTIER S., FEIGENBAUM A., MORTREUIL P., REYNIER A., DOLE P., RIQUET A.M. 1998 Interaction of a vinylic organosol used as can coating with solvents and food simulants, J. Agric. Food Chem. In press

(9)   HAMDANI M., FEIGENBAUM A. 1996: Migration from plasticised poly(vinyl chloride) into fatty media: importance of simulant selectivity for the choice of volatile fatty simulants; Food Additives and Contaminants, **13**, 717-730

(10)   FIGGE K.,1980, Migration of components from Plastics-Packaging Materials into Packed Goods- Test methods and diffusion models. Prog. Polymer Science, **6**, 187-252

(11)   CHANG S.S., SENICH G.A. SMITH L.E., 1982, Migration of low molecular additives in polyolefines and copolymers. Food and Drug Administration Washington DC.

(12)   DE KRUIJF N., RIJK M. A. H. 1997. The suitability of alternative fatty food simulants for overall migration testing under both low and high temperature test conditions. Food Additives and Contaminants,**14**, 775-789.

(13)   VAN KREVELEN D. W., 1990: Properties of polymers. Their correlation with chemical structure; their numerical estimation and prediction from additive group contributions. Elsevier, 3rd Edition, Amsterdam pp. 189-225

(14)   ALLAIN 1993: Migration globale et simulants solvants; Alimentarité des matières plastiques et des caoutchoucs, International Symposium, Carré des Sciences, Paris, 1-2 avril 1993

(15)   DUCRUET V.J., RASSE A., FEIGENBAUM A.E., 1996: Food and packaging interactions: use of Methyl Red as a probe for PVC swelling by fatty acid esters. Journal of Applied Polymer Science, **62**, 1745-1752

(16)   RIQUET, A.M., HAMDANI M., FEIGENBAUM A., 1995/96: Interaction between PVC and simulator media, using paramagnetic probes. Journal of Polymer Engineering, **15**, 1-16

(17)   RIQUET A.M., FEIGENBAUM A., 1997: Food and Packaging Interactions: tailoring fatty food simulants; Food Additives and Contaminants, **14**, 53-63

(18)   RIQUET A. M., VERGNAUD J. M., FEIGENBAUM A.E., 1998: Food Additives and Contaminants, Accepted.

(19)   HAMDANI M., FEIGENBAUM A., VERGNAUD J. M. 1997: Prediction of worst case migration from packaging to food using mathematical models; Food Additives and Contaminants, **14**, 499-506

Chapter 8

What Simulant Is Right for My Intended End Use?

Melanie McCort-Tipton and Robert L. Pesselman

Covance Laboratories, 3301 Kinsman Boulevard, Madison, WI 53704

Choosing the right simulant for a food packaging migration test can be challenging. Making the correct choice depends on several factors including the material being tested, the food products with which the material will be in contact, and the governing body to which the data will be submitted. This paper will present some general guidelines to follow in choosing the correct simulant.

Tests to determine the migration of indirect additives into actual foods are often difficult because the food products themselves interfere with the tests. As a result, tests are done using food-simulating solvents which mimic the leaching action of aqueous, acidic, alcoholic, and fatty foods. In order to determine the appropriate food simulant to be used in testing, several questions must be answered.

- Where will the product be marketed?
- Is the product regulated (compliance vs. migration testing)?
- What types of food will be in contact with the product?
- Can the number of simulants be reduced?
- Can alternate or substitute simulants be used, if so when?

The first step is to determine in what countries the product will be marketed. This will have a direct effect on the simulants allowable for testing. In this paper, the focus is on United States (US) and European Union (EU) requirements.

After identifying the marketplace, it must be determined as to whether or not the product is regulated. If a material is regulated in the US, then the required simulants are specified in the applicable regulations. Generally, these simulants are water, n-heptane, and 8% ethanol. The particular simulant used is dependent on the food type that will be in contact with the product. If the material is not regulated, then the simulants are defined in the Food and Drug Administration's (FDA), "Recommendations for Chemistry Data for Indirect Food Additives Petitions"[1]. The simulants for regulated and unregulated materials in the EU are the same and are defined in Commission Directive 97/48/EC.[2]

The specific food types will dictate the simulant to be used for migration testing. However, it may not be necessary to conduct the migration tests with every simulant that is applicable. In addition, there are exceptions to the recommendations that require the use of alternate or substitute simulants.

FDA Recommendations

The FDA recommends simulants based on the food type. These food types, as defined in *the Code of Federal Regulations*[3] are shown in Table I.

Table I. FDA Classification of Types of Raw and Processed Foods

1. Nonacid, aqueous products; may contain salt or sugar or both (pH above 5.0). **[Aqueous]**
2. Acid, aqueous products; may contain salt or sugar or both, and including oil-in-water emulsions of low- or high-fat content. **[Acidic]**
3. Aqueous, acid or nonacid products containing free oil or fat; may contain salt, and including water-in-oil emulsions of low- or high- fat content. **[Fatty]**
4. Dairy products and modifications:
   A. Water-in-oil emulsions, high- or low-fat. **[Fatty]**
   B. Oil-in-water emulsions, high- or low-fat. **[Aqueous]**
5. Low-moisture fats and oils. **[Fatty]**
6. Beverages:
   A. Containing up to 8 percent alcohol. **[Alcoholic]**
   B. Nonalcoholic. **[Aqueous]**
   C. Containing more than 8 percent alcohol. **[Alcoholic]**
7. Bakery products other than those included under Types VIII or IX of this table:
   A. Moist bakery products with surface containing free fat or oil. **[Fatty]**
   B. Moist bakery products with surface containing no free fat or oil. **[Aqueous]**
8. Dry solids with the surface containing no free fat or oil. **[None]**
9. Dry solids with the surface containing free fat or oil. **[Fatty]**

Because the fat content and pH are important factors in migration, the correct simulant must be chosen for each test. The classic simulants include 3% acetic acid for acidic foods, heptane or corn oil for foods with fat content greater than 5%, and water/ethanol solutions. Recently the FDA has recommended the use of ethanol solutions instead of traditional simulants. The FDA recommended simulants for each food type are shown in Table II.

Table II. FDA Recommended Food Simulants

Food Type	Food Simulant
Aqueous and Acidic	10% Ethanol
Alcoholic	10%[a] or 50% Ethanol
Fatty	Food oil, HB307[b], or Miglyol 812[c]
Dry None	

[a] 10% ethanol can be used for foods up to 15% alcohol content.
[b] A mixture of synthetic triglycerides, primarily C_{10}, C_{12}, and C_{14}.
[c] A fractionated coconut oil.

When more than one food type will be in contact with the material of interest, the recommended simulant for each food type should be used for the migration testing. In most cases, 10% ethanol can be used to simulate aqueous and acidic foods. However, there are some exceptions:

- If the acidity of the food to be in contact with the material is expected to increase the migration of the component of interest, the material should be extracted with both water and 3% acetic acid.
- If the polymer or adjuvant is acid sensitive, the material should be extracted with both water and 3% acetic acid.
- If trans esterification occurs in ethanol, the material should be extracted with both water and 3% acetic acid.

The FDA allows the use of several different simulants for evaluating migration into fatty foods. If possible, a food oil is typically the best choice. However, if the extractions are to be conducted at higher temperatures, then the oxidation of a food oil, such as corn oil, might pose analytical problems. Therefore, HB307 or Miglyol 812 may be better simulants for high temperature applications.

In addition to the above recommended fatty food simulants, the FDA allows the use of aqueous-ethanol simulants. These can be substituted when the use of a food oil (e.g., corn or olive oil) is not practical. As shown in Table III, the specific simulant is dependent on the polymer being tested.

Table III. FDA Recommended Food Oil Substitute Simulants

Polymer	Food Simulant
Polyolefins	Absolute or 95% ethanol
Ethylene-vinyl acetate copolymers	Absolute or 95% ethanol
Rigid polyvinyl chloride	50% ethanol
Polystyrene	50% ethanol
Rubber-modified polystyrene	50% ethanol

In situations where a material is being tested for use with high alcoholic foods using an ethanolic simulant for the fatty food simulant. Instead, a "worst-case" simulant can be used to represent both types of food.

In addition, the FDA still allows the use of heptane as a fatty food simulant in specific cases. It can only be used when very low migration is expected, such as for an inorganic adjuvant or a highly cross-linked polymer. However, correction factors can no longer be applied unless there is sufficient evidence to justify their use. The FDA's alternate recommended food simulants for each food type are:

Food Type	Alternative Food Simulant
Aqueous and Acidic	Distilled water and 3% acetic acid
Alcoholic	None
Fatty	50% or 95% ethanol or heptane
Dry	None

European Union

The EU's recommended food simulating solvents are contained in the Commission Directive 97/48/EC.[4] Like the FDA, the EU has classified the simulants by food type. The EU classifications are presented in Table IV.

Table IV. European Union Food Type Classifications

Food Type	Food Simulant	Abbreviation
Aqueous (pH>4.5)	Distilled water	Simulant A
Acidic (pH<4.5)	3% Acetic acid	Simulant B
Alcoholic	10% Ethanol	Simulant C
Fatty mixture of triglycerides, Sunflower oil, or Corn oil[a]	Rectified olive oil, synthetic	Simulant D
Dry	None	None

[a]The corn oil must meet certain specifications.

A list of foods and the appropriate food simulants are presented in Table V. In addition, the EU has also provided a list of recommended simulants when more than one food type will be in contact with the material of interest. In those cases, testing is not required with all applicable simulants. These food type combinations and the recommended simulants are:

Food Types	Simulant
Aqueous and acidic	Simulant B
Aqueous and alcoholic	Simulant C
Aqueous, alcoholic, and fatty	Simulants C and D

A substitute fatty simulant may be used, provided it can be demonstrated that the use of simulant D in not technically practical. In addition, rather than using one simulant, three simulants must be used. The substitute simulants are isooctane, 95% ethanol, and modified polyphenylene oxide. The modified polyphenylene oxide testing is omitted for certain test conditions (i.e., 10 day at 5°C, 10 days at 20°C, 10 days at 40°C, or 2 Hours at 70°C)

The use of all three substitutes is not required when there is scientific evidence to support that the use of the simulant is not appropriate for the material of interest. In addition, if the material of interest undergoes unexpected changes during the extraction with a substitute simulant, the data derived from that simulant may be discarded.

An alternative fat test may be used in conjunction with other simulants, provided the following conditions are met:

* The results are equal to or greater than those obtained using the simulant D.
* The migration does not exceed the migration limits after application of an appropriate correction factor, which can be found in Directive 85/572/EEC.

Table V: EU Recommended Simulants for Various Foods

Description of Foodstuffs	Simulants
Non-alcoholic beverages or alcoholic beverages with <5% (v) alcohol	A or B[1]
Alcoholic beverages ≥5% (v) alcohol	B[2] and C[3]
Miscellaneous: undenatured ethyl alcohol	B[2] and C[3]
Pastry, biscuits, cakes, and other baker's wares, dry: with fatty substances on the surface	
A. Dry: with fatty substances on the surface	D
B. Fresh: with fatty substances on the surface	D
C. Fresh: other	A
Chocolate, chocolate-coated products, substitutes and products coated with substitutes	D
Confectionery Products	
A. In solid form: with fatty substances on the surface	D
B. In paste form: with fatty substances on the surface	D
C. In paste form: moist	A
Sugar and sugar products	
A. Honey and the like	A
B. Molasses and sugar syrups	A
Processed fruit	
A. Fruit in the form of chucks, puree or paste	A or B[1]
B. Fruit preserves	
1. In an aqueous medium	A or B[1]
2. In an oily medium	A or B[1], and D
3. In an alcoholic medium (≥5% vol.)	B[2] and C
Nuts	
A. Shelled and roasted	D[4]
B. In paste or cream form	A and D[4]
Processed vegetables:	
A. Vegetables, cut, in the form of purees	A or B[1]
B. Preserved vegetables:	
1. In an aqueous medium	A or B[1]
2. In an oily medium	A or B[1], and D
3. In an alcoholic medium (≥5% vol.)	B[2] and C
Animals and vegetable fats and oils, whether natural or treated	D
Margarine, butter and other fats and oils made from water emulsions in oil	D

[1] If the pH is ≤4.5 then use simulant B, otherwise use simulant A.
[2] Use simulant B only if the pH is ≤4.5.
[3] If the alcohol content is >15% vol. ethanolic simulants can be substituted.
[4] If it can be showed that there is no "fatty contact" then simulant D can be omitted.

Continued on next page.

Table V (continued)

Description of Foodstuffs	Simulants
Fish:	
A. Fresh, chilled, salted, smoked	A and D[4]
B. In the form of paste	A and D[4]
Crustaceans and mollusks (including oysters, mussels, snails) not naturally protected by their shells	A
Meat of all zoological species (including poultry and game):	
A. Fresh, chilled, salted, smoked	A and D
B. In the form of paste, creams	A and D
Processed meat products (ham, salami, bacon, and other)	A and D
Preserved and part-preserved meat and fish:	
A. In an aqueous medium	A or B[1]
B. In an oily medium	A or B[1], and D
Eggs not in shell other than powdered or dried	A
Liquid egg yolks	A
Milk:	
A. Whole	A
B. Partly dried	A
C. Skimmed or partly skimmed	A
Fermented milk or yogurt, buttermilk, and such products in association with fruit and fruit products	B
Cream and sour cream	A or B[1]
Cheeses:	
A. Processed cheeses	A or B[1]
B. All others (except whole, with rind)	A or B[1], and D[4]
Fried or roasted foods:	
A. Fried potatoes, fritters and the like	D
B. Of animal origin	D
Preparations for soups, brothes, in liquid, solid or powder form; homogenized composite food preparations, prepared dishes:	
A. Powdered or dried: with fatty substances on the surface	
B. Liquid or paste:	
1. With fatty substances on the surface	A or B[1], and D
2. Other	A or B[1]
Yeast and raising agents: in paste form	A or B[1]

[1] If the pH is ≤4.5 then use simulant B, otherwise use simulant A.
[2] Use simulant B only if the pH is ≤4.5.
[3] If the alcohol content is >15% vol. ethanolic simulants can be substituted.
[4] If it can be showed that there is no "fatty contact" then simulant D can be omitted.

Table V (continued)

Description of Foodstuffs	Simulants
Sauces:	
A. Without fatty substances on the surface	A or B[1]
B. Mayonnaise, sauces derived from mayonnaise, salad creams and other oil in water emulsions	A or B[1], and D
C. Sauce containing oil and water forming two distinct layers	A or B[1], and D
Mustard (except in natural state)	A or B[1], and D[4]
Sandwiches, toasted bread and the like containing any kind of foodstuffs: with fatty substances on the surface	D
Ice-creams	A
Dried foods: with fatty substances on the surface	D
Concentrated extracts of an alcoholic strength ≥5% vol	B[2] and C
Cocoa:	
A. Cocoa powder	D[4]
B. Cocoa paste	D[4]
Liquid coffee extracts	A

[1]If the pH is ≤4.5 then use simulant B, otherwise use simulant A.
[2]Use simulant B only if the pH is ≤4.5.
[3]If the alcohol content is >15% vol. ethanolic simulants can be substituted.
[4]If it can be showed that there is no "fatty contact" then simulant D can be omitted.

Table VI: EU Recommended Fatty Food Simulants

Recommended Food Type	Substitute Food Simulant	Food Simulant
Aqueous (pH>4.5)	Distilled water	None
Acidic (pH<4.5)	3% Acetic acid	None
Alcoholic	10% Ethanol	None
Fatty	Rectified olive oil, synthetic mixture of triglycerides, sunflower oil, or corn oil	Isooctane, 95% ethanol, and modified polyphenylene oxide
Dry	None	None

The EU defines the alternative fat tests as follows.

- Alternative tests with volatile media: These tests are conducted with volatile simulants, such as isooctane or 95% ethanol, at conditions such that the results are greater than or equal to those which would be obtained using simulant D.
- Extraction tests: These tests are conducted using other media which, based on scientific data, demonstrate strong extraction power under severe conditions. The results must be equal to or greater than those that would result from the use of simulant D.

The EU's recommended fatty food simulants are summarized in Table 6.

If data is to be submitted to more than one country (e.g., the US and the EU), it is recommended that the "worst-case" simulant for each food type be used.

Food Type	FDA Simulant	EU Simulant
Aqueous (pH >4.5)	10% Ethanol	Distilled water
Acidic (pH ≤4.5)	10% Ethanol	3% Acetic acid
Alcoholic	10% or 50% Ethanol	10% Ethanol
Fatty	Food oil	Rectified olive oil

Conclusion

In general, the use of the recommended simulants for aqueous, acidic, and alcoholic foods present no problems. Most problems associated with migration studies are due to problems that arise from the use of the recommended fatty food simulants. This is due to the fact that in many cases the analytes of interest are lipophilic, making the analysis in food oils difficult. The FDA allows for the use of ethanolic simulants, when the traditionally recommended simulants are not practical. The EU allows for alternate choices, but only after demonstrating that a suitable method could not be established with the recommended simulants or if there is scientific evidence showing that an alternative provides results that are equal to or greater than those obtained from simulant D.

Literature Cited

1. Center for Food Safety and Applied Nutrition, "Recommendations for Chemistry Data for Indirect Food Additive Petitions." Food and Drug Administration, Washington, D.C., 1995, pp.
2. European Economic Community (EEC), "Commission Directive No. 85/572/EEC of 19 December 1985 laying down the list of simulants to be used for testing migration of constituents of plastic materials and articles intended to come in contact with foodstuffs." *Official Journal of European Communities*, L 372, 31 December 1985, p. 14.
3. Federal Register, "21 CFR 176.170 Components of paper and paperboard in contact with aqueous and fatty foods." *Code of Federal Regulations 21 Parts 170 – 199*, U.S. Government Printing Office, Washington, D.C., 1998, pg. 201.
4. EEC, "Commission Directive No. 97/48/EC of 29 July 1997 amending for the second time Council Directive 82/711/EEC laying down the basic rules necessary for testing migration of the constituents of plastic materials and articles intended to come in contact with foodstuffs." *Official Journal of European Communities*, L 222, 8 December 1997, p. 10.

Chapter 9

Stability of R-PET for Food Contact: Pilot Recycling Studies and Mathematical Confirmation

George Sadler[1], S. Hussaini[1], and D. Pierce[2]

[1]Illinois Institute of Technology, National Center for Food Safety and Technology, South Archer Road, Summit-Argo, IL 60501–1933
[2]William Paterson University, Wayne, NJ 07470

The ability of mathematical equations to predict surrogate uptake at various stages in contamination and recycling processes was examined. Surrogate uptake by PET flakes bathed in a surrogate cocktail was greater than predicted by mathematical models for most compounds. Contradiction between actual and calculated uptake appeared to arise from capillary adsorption on grinder-damaged flake. Desorption predictions were also influenced by this surface-retained reservoir of surrogates. However, once surrogates were impartially mixed through the polymer mass by diffusion, mathematical calculations provided a good prediction of migrations into foods.

Dozens of studies have been published on the mathematics of diffusion in an effort to better understand the fate of contaminants and adjuvants held inside polymers. By contrast, relatively little information is available on the quality of prediction mathematical models provide for each treatment step in a typical recycling study. The United States Food and Drug Association (FDA) has proposed a surrogate challenge approach to evaluate whether polymers tainted through consumer recycling abuses are safe for remanufacture into new containers once they have been treated by recycles to remove contaminants. Mathematical models would provide a valuable predicative tool if they could determine, *a priori,* the absorption of compounds from a surrounding surrogate medium, or if they could accurately model the release of trapped contaminants from a second generation container. Reliable models would also be valuable in assessing the effectiveness of novel cleaning treatments or suggesting optimizations to existing treatments.

The purpose of this study was to compare mathematical predictions of surrogate behavior with actual data in an FDA "Points to Consider" type recycling run with poly(ethylene terephthalate), (PET) . The treatment steps investigated included were: surrogate absorption from a surrogate cocktail and removal through washing, drying and extrusion.

Materials and Methods

Surrogate Contamination of PET. Polymer (flaked virgin PET soda bottles) batches of 100-200 Kg were contaminated for 2 weeks at 40˚C with a surrogate solution containing representatives from all FDA volatility and polarity classes. The cocktail contained: toluene (10%, non-polar, volatile), chloroform (10%, polar-volatile), lindane (1%, non-polar, non-volatile), benzophenone (1%, polar, non-volatile) and CuII 2-ethylhexanoate (1%, heavy metal). Tectracosane (1%) was used as a second non-polar, non-volatile surrogate with extremely low volatitility. Flake was packed (1600-1800 g/container) into 4 L glass containers and filled with surrogate solution to within 5 cm of top polymer level. Containers were incubated for 2 weeks at 40˚C. Thermal expansion of the surrogate solution during incubation covered the remaining polymer.

Alternately, 2 L of surrogate solution was filled into whole virgin PET bottles. Bottles were then capped with standard closures and incubated at 40˚C for 2 weeks.

Washing PET Flake. At the end of storage, the PET/surrogate slurry was poured from gallon containers into a 400 L steam-jacketed reaction vessel with a "lightning mixer" agitator. The slurry was allowed to drain for 30 minutes through a screen fitted at the bottom of the vessel. Afterward, the drain was shut and the vessel was filled with cold tap water to approximately 6 cm above the polymer surface (approximately 150-240L water). A Okite R28 surfactant was added to a 1.5% final concentration in the wash water. The PET/water/Okite slurry was then heated to 92˚C over approximately 15 minutes. The temperature was held at 92˚C for 15 minutes and then drained. Following a 15 minute drainage period, the drain was closed, and the vessel refilled with cold water to approximately 6 cm above the surface of the flake. The mixer was used to suspend the flake in slurry for 15 minutes after which the valve was reopened and the water allowed to drain for 15 minutes. Flake was removed manually from the washing vessel in approximately 11-14 kg batches and spun in a basket centrifuge to remove bulk water.

Washing Whole PET Bottles. Bottles were cleaned as described above for flake except drained whole bottles were transferred to cold water containing 1.5% Okite R28 surfactant. Floaters were submerged to fill with wash solution. Some remained slightly buoyant. As previously described, the vessel was brought to 92˚C, drained, refilled with cold water, agitated and drained again. Bottles were removed manually and the remaining solution was drained from the individual bottles. The bottles were ground in a centrifugal grinder over a 22 mesh screen.

Drying of Flake. Centrifuged flakes were dried for 4 hours at 160˚C to a final dew point of -120˚C. Dried flake was immediately transferred to plastic bags and closed with a twist tie to inhibit moisture reabsorption.

Extrusion. PET was extruded into 25 mil ± 2 mil sheet (1 mm ± .08) in a Haake single screw extruder with (15 ± 1 inHg abs) or without vacuum at the vent port.

Polymer Analysis

Total Dissolution (PET) . Total uptake of all surrogates except Cu^{+2} 2-ethylhexanoate were determined through total dissolution of PET with trifluoroacetic acid (TFA). $2 \pm$ .001 g of PET was treated with 15 mL of trifloroacetic acid (TFA). The combination was agitated on a wrist action shaker until the flake was fully dissolved. The digest was extracted 3 times with 15 mL of heptane. The extract was washed 2 times with 10 mL of water to remove residual TFA. The washed heptane layer was collected and diluted to 50 mL with neat heptane. 1 µL of the heptane sample was analyzed chromatographically using GC/MS. Tests were performed in triplicate.

Cu^{+2} 2-ethylhexanoate was determined through ashing of the polymer at $500°C$ until carbonacious matter was removed. The light ash was dissolved in 0.1 N HNO_3 and Copper was quantified using atomic absorption for copper residuals of > 0.5 ppm or stripping anodic voltammetry for copper residuals < 0.5 ppm.

Food Simulating Solvent (FSS) Extraction . Disks,1.7 cm in diameter, were punched from extruded sheet using an arch punch. Disks (n=25) were strung on chromel wire using glass bead spacers following the method of Snyder and Breder (1985)[1]. The string of disks were placed in an EPA vial which was filled to a bulging meniscus with either heptane (fatty FSS) or 10% ethanol (aqueous FSS). 1 µL were examined by GC/MS at 2, 24, 96, and 240 hours. Tests were conducted in triplicate.

Chromatographic Conditions. Chromatography was performed on a Hewlett Packard 5890 Series II GC with a 5971 mass selective detector using select ion monitoring mode (SIMM). Separation was achieved with a 30 m, 0.32 mm ID, DB-1 column (J&W Scientific, Folsom, CA).

Diffusion Coefficient. Surrogate diffusion coefficients were determined by one of two methods. Diffusion coefficients of volatile compounds were measured with a MAS 2000 organic vapor permeation unit (MAS Technologies, Umbrota MN). Volatile vapors were generated in a thermostated ($25°C$) sparger. GC comparison of sparger vapor with headspace over neat volatile at $25°C$ confirmed that the sparge stream was saturated with the test vapor. Diffusion was determined from a modification of Pasternak, (1970)[2] through the equation:

$$D = (slope_{mas} \cdot \ell^2)/ (5.92 \cdot \textbf{Permeation}_{eq}) \qquad (1)$$

Where ℓ is the polymer thickness and Permeation$_{eq}$ is the equilibrium permeation.

Diffusion coefficients of non-volatile compounds were determined after the method of Sadler et al., (1996)[3] where 30% solutions of non-volatile compounds were made in water, ethanol or heptane and sealed into ~25 cm^2 packets of the test polymer. Packets were placed in 30 mL glass vials fitted with mininert sampling closures. Vials were filled with ethanol or heptane to a bulging meniscus to exclude all air. The medium external to the packet was sampled chromatographically at various times or subjected to atomic absorption analysis in the case of CuII 2-ethylhexanoate. Diffusion values were calculated from the relationship:

$$D = \ell^2/6\theta \qquad (2)$$

Where θ is the x-intercept (seconds) of the steady state portion of the diffusion line. When diffusion interpretation was ambiguous due to data scatter, analytical solutions (Eq. 3) for various diffusion values were compared with collected data and the best fit of the non-steady state portion of the line was selected as the reported diffusion value.

$$\frac{Q}{lC_1} = \frac{Dt}{l^2} - \frac{1}{6} - \frac{2}{\pi^2}\sum_{k=0}^{\infty}\frac{(-1)^n}{n^2}\exp^{(-Dn^2\pi^2\frac{t}{l^2})} \qquad (3)$$

Where D is the diffusion coefficient, l is the polymer thickness, t is time, Q is the total permeation at time t and C_1 is the maximum solubility in the polymer under the conditions of the test.

Surrogate Absorption Model. Surrogate absorption was assumed to be Fickian. In order to minimize solvation, no surrogate was present above 10% in the cocktail. Surrogates were dissolved in heptane which was not monitored as a surrogate. Heptane had been shown in previous screening trials to have minimal interaction with PET. An absorption model was used which describes absorption at the surface of a plane film initially free of migrating compound (Crank, 1975).[4]

$$M_t/M_\infty = \sum_{n=0}^{\infty}\frac{8}{(2n+1)^2\Pi^2}\exp^{-Dt\Pi^2(2n+1)^2/4l^2} \qquad (4)$$

and desorption

$$M_t/M_\infty = 1 - absorption \qquad (5)$$

All variables are as previously described. However, when absorption/desorption occured from both sides of the polymer, l was half the polymer thickness. If exchange was from a single side, l was the full polymer thickness.

Results

Six PET recycling runs for 4 separate companies have been undertaking over the last 2 years. All recycling steps were not conducted in each study. Some involved only contamination. Others contamination and cursory cleaning. Two studies included all steps from contamination to extrusion including subsequent extraction with FSS.

Although, surrogate behavior was similar in direction for each test, initial surrogate loads often differed notably in amount.   In general, the relative standard error (RSE) for volatile surrogate uptake (determined by total dissolution) was as great as ± 100% between studies.   The RSE for non-volatile uptake could be as great as ± 400%. Differences in PET manufacturers and flake size were uncontrolled and were likely responsible for some of the differences observed. Lot-to-lot variability in surrogate solution may also have accounted for some differences. The data reported in this paper are largely from a single all-processing-step study. However, interesting findings from other data sets are discussed where appropriate.

Surrogate Uptake at 40°C for 2 Week Incubation. The diffusion coefficients and solubility constants for various surrogates in PET (40°C) are given in Table 1 along with calculated and actual uptakes following incubation for 2 weeks at 40°C. Calculated uptakes assumed that at the 1 to 10% surrogate concentrations used in this study that chemical activity and chemical concentration were the same. Surrogate uptakes were calculate from the equation 4 and stated in terms of ppm through the relationship:

$$ppm = M_{14}/M_\infty \cdot \frac{S}{\rho} \qquad (6)$$

Where M_{14}/M_∞ is the percent of saturation at 14 day for the diffusion coefficient of the permeant in the polymer at 40°C, calculated from equation 4. The value ρ is the polymer density. S is the solubility constant which was calculated by the expression:

$$S = S_{ext} \cdot \frac{f \cdot V_{40}}{V_T} \qquad (7)$$

S_{ext} was the extrapolated surrogate solubility at 40°C derived from a series of runs at higher temperatures. V_{40} was the vapor partial pressure of the permeant at 40°C and V_T was the pressure of pure permeant vapor pressure at the condition of the run taken from existing tables[5]. The variable f is the volume fraction of the specific surrogate in the cocktail.   Consequently, under ideal conditions, S would be the product of the volume fraction of the surrogate in the solution and the solubility of neat permeant or its saturated vapor.

Coefficients, estimated and actual surrogate uptakes are given below for migration into a .0381 cm (15 mil) PET soft drink container.

Basis for Poor Calculated Surrogate Absorption Values

Migration Coefficients as a Possible Source of Error. In every case, surrogate uptake greatly exceeded the calculated value.   Actual values for non-volatile surrogates exceeded calculated values by several thousand fold. Some discrepancy was inevitably due to issues of departure from the stated assumption of ideality. Compounds characteristically exhibit a positive departures from Raoults law when dissolved in a sparingly soluble menstrum.[6] The surrogate solution was barely miscible due largely to the use of isopropyl alcohol as a solvent for CuII, 2-ethylhexanoate and therefore may have set up an environment where activities were greater than their volume fraction would predict. Since chloroform and toluene begin to plasticize PET above a vapor activity of 0.1, an activity greater than the volume fraction could have plasticized the polymer allowing opportunistic absorption of all surrogates.

Extrapolation of solubilities from high temperature could have also have potentially accounted for part of the error. PET migration tests were usually conducted between 120-160°C. Extrapolations to 40°C did not take into account Arrhenius non-linearity in diffusion and solubility which likely occur at the glass transtion (t_g) of PET (73-80°C). Below the glass transition, arrhenius lines do usually become less steep. Although a simple extrapolation which ignores the glass transition would tend to calculate values which would underestimate migration, such inaccuracies would at most be limited to a few fold and not orders-of-magnitude.

The surrogate solution was formulated in heptane as an inert solvent (66% by volume). Screening experiments with heptane suggested that it had very little interaction with PET. However, some solvation with subsequent opportunistic absorption of other components cannot be totally dismissed.

Differences in the physical properties of the PET film used to determine coefficients and PET flake used for contamination studied may also have introduced some inaccuracy. Coefficients in Table 1 were determined on a .00127 cm (1/2 mil) PET film with 50% crystallinity. The crystallinity of PET flake lies in a continuum from nearly 0 to approximately 50%. Therefore, absorption predictions in Table 1 based on a crystalline film could be low for mixed crystallinity flake by as much as 2 fold.

Cumulatively, these sources of error could only produce discrepancies observed in Table 1 if they represented orders-of-magnitude departures from ideality. The relative inertness of heptane on PET and restricting surrogate components to ≤ 10% of cocktail concentration, argues against such vast deviations. Certainly, they cannot account for the uptake levels for CuIIETOH, a compound which Rutherford backscattering indicates is unabsorbed from the surrogate cocktail[7]

Surface Anomalies as a Potential Source of Absorption Error. Clearly some surrogate will be associated with the surface of the polymer through adsorption. If the excess surrogate uptake values observed in Table 1 are due to this surface phenomena, then whole bottle absorption should be approximately half the value for flake absorption. Table 2 lists the results of a whole bottle surrogate contamination run.

Whole bottle values were much lower than half of flake values. Similar differences were observed by Komolprasert, et al, (1994).[5] This disparity suggests that excessive surrogate uptake by flake was due to issues other than simple surface absorption.

Cold water washing precipitated a waxy film in the wash water. It appears that cracks, stress fissures and feathered edges generated during flaking provided capillary sinks for surrogate adsorption. The precipitated waxy film essentially cemented trapped

Table 1. Calculated and Actual Surrogate Uptake by PET Flake Treated for 2 Weeks at 40°C

Surrogate	Diffusion (40°C) cm^2/sec	Solubility (40°C) Saturated Vapor, g/cm^3	Calculated Uptake M_t/M_∞[a]	ppm	Actual Uptake (ppm)[b]
Chloroform (10%)	2.28×10^{-13}	.00320	31×10^{-3}	72	$954 \pm 1.3\%$
Toluene (10%)	2.46×10^{-13}	.00043	32×10^{-3}	7.2	$4105 \pm 7.4\%$
Benzophenone (1%)	1.70×10^{-14}	.000025	8.5×10^{-3}	0.15	$257 \pm 68\%$
Lindane (1%)	3.50×10^{-15}	.000032	4.0×10^{-3}	0.10	$532 \pm 50\%$
Tetracosane (1%)	5.0×10^{-16}	$\approx .000005$	2.3×10^{-3}	0.01	$255 \pm 57\%$
Cu II, 2-ethylhexanoate (1%)	$<1 \times 10^{-16}$	none[c]	0	0.0	$140 \pm 78\%$

[a] M_t/M_∞ values were confirmed by numerical treatment

[b] Following cold water wash with 1.5% Okite R-28 surfactant

[c] Confirmed by Rutherford Backscattering

surrogates inside these irregular flake features. This surface retention model could explain the excessive surrogate uptake apparent in Table 1. It is apparent from Tables 1 and 2 that uptake of all surrogates from a contaminant cocktail is complex and poorly explained by any straightforward model of ideal diffusion.

Surrogate Loss During Drying

In industrial drying of PET flake, heated air is continuously re-circulated through desiccant driers. However, the desiccant bed which removes water from recirculating air in commercial processing is not able to fully absorb volatilized surrogates. As a result, the re-circulated drying air quickly becomes saturated with desorbed volatiles. To prevent this problem, re-circulated air and desiccant resins were not used in drying experiments. Flake was not stirred during drying.

Numerical evaluation was used to calculate residual surrogate concentration in surrogate-exposed flake after drying at 160°C for 4 hours. Calculations assumed that "actual uptake" values from Table 1 were correct and represented Fickian absorption into the polymer via Eq 4, a conclusion which, as previously discussed, is refuted by several lines of reasoning. Consequently, even though calculated residuals appear to be in the correct order-of-magnitude for several surrogates, correct predictions may be lucky guesses in which diffusion coefficients fortuitously model complex phenomena partially or totally unrelated to surrogate migration within the polymer. This premise can be tested.

If, as previously postulated, excessive surrogate was localized on the polymer surface prior to drying then "calculated" residual losses would be greater than "observed" residual losses. This follows since surface-retained surrogates are spared the nuisance of slow diffusion through the polymer matrix and should therefore be liberated with greater ease than ideal polymer diffusion models would predict. Tetracosane and CuII clearly demonstrated this pattern. The pattern was present; but much less pronounced for chloroform and toluene. The absence of this tendency for benzophenone and lindane may simply reflect a bad prediction by the diffusion value for these compounds. Had the resin bed been stirred, actual residuals might have been lower in every case. Consequently, while some surrogates losses (volatiles in particular) invite hope that ideal models can predict surrogate behavior during drying, until the complex impact of surface phenomena can be understood, the predictive competence of equation drying models must be considered inconclusive.

Surrogate Losses Via Food Simulating Extraction

Resin from Table 3 was subjected to extrusion in an unvented extruder. It was assumed that the "actual residual" concentration of all surrogates were impartially mixed throughout the extruded polymer without significant loss. FSS extraction results are given in tables 4 and 5 for fatty and aqueous FSS respectively.

Diffusion equations correctly predicted that surrogate extraction by FSS would be near the detection limit or below for each of the surrogates. Predicted values for chloroform were low. Continental PET Technologies used chloroform as the volatile-polar surrogate in their submission to FDA for multilayer PET soda bottles. The extraction of chloroform was also greater in that study than for any other surrogates. Lower predicted values reported in Table 4 may indicate the stated diffusion coefficient

Table 2. Surrogate Uptake by Whole PET Bottles Treated for 2 Weeks at 40˚C

Surrogate	Actual Uptake (ppm)[a]
Chloroform (10%)	323 ± 22%
Toluene (10%)	161 ± 33%
Benzophenone (1%)	35 ± 57%
Lindane (1%)	40 ± 15%
Tetracosane (1%)	39 ± 74%
Cu II, 2-ethylhexanoate (1%)	21± 141%

[a] Following cold water wash with 1.5% Okite R-28 surfactant

Table 3. Actual and Calculated Surrogate Residuals
in PET Flake Following Drying for 4 hours at 160˚C

Surrogate	Diffusion at 160˚C cm^2/sec	Calculated Residual Post -Drying (ppm)	Actual Residual Post-Drying (ppm)
Chloroform (10%)	1.22×10^{-8}	18	13
Toluene (10%)	4.88×10^{-9}	160	75
Benzophenone (1%)	3.72×10^{-9}	5	17
Lindane (1%)	8.1×10^{-11}	68	83
Tetracosane (1%)	4.1×10^{-14}	252	11
Cu II, 2-ethylhexanoate (1%)	0	140	4

Table 4. Calculated and Actual Surrogate Losses by PET Subjected to Extraction by Hexane and 10% Ethanol Food Simulating Solvents

Surrogate	Diffusion (40˚C) cm^2/sec	Calculated desorption ppb	Actual Desorption hexane (ppb)	Actual Desorption 10% ethanol (ppb)
Chloroform (10%)	2.28 x 10^{-13}	5.83	31	trace[b]
Toluene (10%)	2.46 x 10^{-13}	33.8	24	trace
Benzophenone (1%)	1.70 x 10^{-14}	3.06	nd[a]	nd
Lindane (1%)	3.50 x 10^{-15}	8.5	nd	nd
Tetracosane (1%)	5.0 x 10^{-16}	0.9	nd	nd
Cu II, 2-ethylhexanoate (1%)	<1 x 10^{-16}	0	nd	nd

[a] approximate 5 ppb and below for most compounds
[b] less than 10 ppb but more than 5 ppb for compounds.

was lower than the true value. Alternately, it could mean that significant losses occurred during extrusion.

Migration equations competently modeled extraction of surrogates by fatty food simulants. Models overestimated extraction by an aqueous food simulating solvent. Solubility and partition issues complicate extraction phenomenon when aqueous solvents are used to extract non-polar compounds. Since equation 4 does not model solvent properties, it will tend to predict the maximum possible extraction rate as long as there is no significant interaction between the polymer and the solvent.

The FDA allows the assumption that only 50% of the recycle stream is contaminated with a single contaminant[8]. The fraction of the diet that comes in contact with PET (FDA's consumption factor) is .05. Under these assumptions, the PET resin examined in this study would impart less than 0.5 ppb of all surrogates to diet if the recycled polymer when used in aqueous food contact. Surrogate extraction by the fatty food simulant exceeded the 0.5 ppb dietary threshold set out by the FDA's Threshold of Regulation Policy. Therefore, PET resin recycled by this process would not be suitable for packaging fatty foods.

It should be emphasized that the recycling process employed in this study do not involve any exceptional cleaning innovations. Equations suggest that modifications such as drying changes, vacuum extrusion or the inclusion of solid stating would bring down values in Table 4 to levels that would allow recycled PET to be used for all food types and at other use temperatures.

Conclusion

Equations appeared to be virtually ineffective in predicting initial uptake of compounds from surrogate solution. Although drying losses appeared correctly modeled for several surrogates, there are unresolved questions concerning whether losses are due to surface evaporation or true diffusion losses or both. In general, chloroform and toluene probably fit the model better than non-volatile compounds which were greatly under-predicted by absorption models. Capillary absorption of surrogate in surface imperfections of the flake was likely responsible to higher than calculated surrogate contamination. Once surrogates were mixed in the polymer through extrusion, mathematical predictions were generally good. However, some departure was observed by chloroform and limits of detection problems deprived the story of a more thorough comparison for FSS extracts. However, in general it appears that the value of mathematical predictions of surrogate behavior during recycling processes improve with each processing step and appear to be better for volatile than non-volatile compounds.

Reference List

1. Snyder, Roger C. and Breder Charles V. *J. Assoc. Off. Anal. Chem.* **1985**, 68(4), 770-775.
2. Pasternak, R. A. Schimscheimer J. F. and Heller J *J. Polymer Sci.*, Part A-2. **1970**, 8, 467-479.

3. Sadler, G. Pierce D. Lawson A. Suvannunt D. and Senthil V. *Food Add. and Contam.* **1996**, 13(8), 979-989.

4. Crank, J. *The Mathematics of Diffusion.* Oxford: Oxford University Press; **1975.**

5. Komolprasert, V. and Lawson A. In. *Plastics, Rubber and Paper Recycling: A Pragmatic Approach.* Washington, D.C.: American Chemical Society; **1995**, 435-444.

6. Weast, Robert. *CRC Handbook of Chemistry and Physics.* Boca Raton, FL: CRC Press; **1985**.

7. Barton, A. M. *CRC handbook of solubility parameters and other cohesion parameters.* Boca Raton, FL, **1983.**

8. Pierce, D. Pfeffer R. and Sadler G . *Nucl. Instr. and Meth. in Phys. Res.* **1997**, 124, 575.

Chapter 10

Recycled Plastics: Experimental Approaches to Regulatory Compliance

Robert L. Pesselman and Melanie McCort-Tipton

Covance Laboratories, 3301 Kinsman Boulevard, Madison, WI 53704

The potential for unknown contaminants in the post-consumer plastic recycled stream has raised safety concerns and resulted in regulatory guidelines for food contact polymers. This paper presents an overview of the current regulatory requirements as well as example study designs and experimental results. Model contamination cocktails and analytical methods are outlined.

The astounding growth and the advancing technologies in the packaging industry have had a dramatic effect on the regulatory and experimental guidelines for determining the impact of packaging on the safety of food. Regulatory and testing concerns are now focused on adapting to meet the ever-changing needs of the marketplace, as more and newer packaging materials are developed and used.

Traditionally, the role of packaging was to prevent contamination. In recent years, new packaging materials and processes have been developed which emphasize convenience for consumers. In addition, making packages tamper-proof has also become a high priority. As a result, packaging has become an integral part of the manufacturing and marketing process. The latest trend in response to consumer demand is to fulfill all of these requirements using biodegradable and recyclable packaging materials.

A major change in the industry is the increasing amount of packaging which includes recycled materials. The safety of using these materials in food packaging is a high priority for the packaging industry and for regulators. The primary concern arising from use of recycled materials is the potential for contaminates to end up in packaged foods.

Laboratory analysis to detect migration of contaminants into packaged foods is an integral part of ensuring safety and quality in food packaging. In the mid-1980s, researchers at testing facilities began working in close association with packaging material manufacturers and food producers in developing methods to ensure the safety of both traditional and recycled materials.

Packaging materials

Food packaging materials such as glass, paper and tin have been largely replaced by aluminum and plastic. Plastic polymers and laminates are used to make everything from bags and pouches, to trays, bottles, and jars with more new uses being developed every year. Of all packaging materials, plastics is the fastest growing segment of the packaging industry. This is due, in large part to its versatility and ease of use. Packaging accounts for almost 30% of the plastic resins produced per year (approximately 19,551,000,000 lb)[1].

New packaging materials pose significant challenges to researchers, because some materials have components that can migrate into the packaged food. This is defined as migration of "indirect food additives" and is regulated in the U.S. by the Food and Drug Administration (FDA). Migration of these indirect additives can cause health concerns and adversely affect the taste, color, and aroma of foods. Researchers' primary goal is ensuring that packaging materials will not pose health hazards through migration of indirect food additives. The complexity of this task has been compounded by the increased growth in the use of recycled materials.

Recycling

Solid waste is a concern for governments worldwide. As more waste gets generated each year, handling methods like landfilling and incineration become more problematic. Plans for reducing solid waste include four methods of management: source reduction (including reuse), recycling, incineration, and landfill.

Packaging materials make up over one-third of the total solid waste generated in the United States.[2] While the amount of solid waste increases, so do recycling efforts. The amount of waste recycled has grown from about 10% in 1980 to an estimated 30% by the year 2000.[3] Some packaging materials are recycled much more than others because of their varying makeup and the nature of recycling technology. The recycling of plastics has grown rapidly. In 1994, over 1 billion pounds of post-consumer plastic bottles were recycled. This is an increase of 21 percent over 1993.[4] This expanding use of recycled materials has resulted in increased attention to ensuring the safety of these materials. When evaluating the safety of packaging materials, especially recycled products, there are three key aspects to be considered: the source of the materials, the nature of the process, and the conditions of use.

Packaging safety regulations

New technologies for improving the quality and lessening the environmental impact of packaging have resulted in more recycled, recyclable, and biodegradable packaging. This has resulted in more complex safety issues, revised regulatory guidelines, and a greater need for testing of packaging materials to ensure that unwanted indirect food additives are not migrating into food products. It is important to understand, however, that the safety concerns pertaining to the use of recycled packages are no different than those established for virgin materials.

Food packaging safety is regulated by the FDA, based on requirements established by the Food, Drug, and Cosmetic Act of 1938 and the Food Additives Amendment of 1958. Any component of packaging that could migrate into the food is considered an indirect food additive and must be approved by the FDA based on premarket safety testing results submitted by the packaging manufacturer.

The regulations and guidelines for approval of packaging materials are continually changing due in part to the rapid increase in the number of uses and mixtures of polymers. In general, the following six factors need to be considered for food packaging to be in compliance.

1. Is the material in contact with the food already regulated by the FDA?
2. Does it contain only additives that are at or below permitted concentrations?
3. To what extent are trace amounts of the constituents of the packaging migrating?
4. What is the total weight of migrating constituents?
5. What is the public health risk of the potential migration?
6. Under what conditions will the packaged product be processed and stored?

In addition, under the provisions of the National Environmental Policy Act of 1968, the FDA is ordered to review the environmental impacts of food packaging materials. In 1995, the FDA adopted a threshold of regulation process for indirect additives. As Bayer[5] noted in his article titled "The threshold of regulation and its application to indirect food additive contaminants in recycled plastics," this threshold "would recognize that there is a level below which the probable exposure to a potentially toxic substance constitutes a negligible risk." He also noted that "This approach, although ultra-conservative, has provided a means for industry and regulators to work together to achieve the goals in reducing municipal solid waste and thereby protect our environment."

If the maximum potential dietary concentration of a substance is below 0.5 parts per billion (ppb) and the substance is not a known carcinogen, the FDA considers it to be safe, based on statistical analysis of the available toxicology data. The dietary concentration is calculated by multiplying the fraction of the food in the diet that is in contact with the packaging material (i.e., the consumption factor), times the average concentration of the additive in food. The material can then be used without having to file a formal indirect food additive petition. If the dietary concentration is above 0.5 ppb a different process must be followed. Filing of an indirect additive petition with the FDA is the traditional method to satisfy this requirement. In these instances, the work of testing laboratories has become very important. In addition to safety, studies must clarify the effect packaging may have on taste, odor, and commercial feasibility.

Migration testing

Migration tests can not only ensure that substances approved as indirect food additives are present at acceptable levels, but identify any components which have not been cleared by the FDA. These tests are required to generate data in support of indirect additive petitions. The major factors to be considered when designing the test methods are the composition of packaged foods, temperatures to which the product and packaging is exposed, the length of time of exposure, and possible migrants such as colorants, plasticizers, and residual monomers.

Determining the migration into actual foods has proven difficult because food products, due to their heterogeneous nature, interfere with the tests. As a result, tests are done using food-simulating solvents which simulate the leaching action of aqueous, acidic, alcoholic, and fatty foods.[6] Because the fat content and pH are important factors in migration, the correct simulant must be chosen for each test.

The high temperatures to which many multipurpose packages are exposed produce favorable conditions for migration. The time and temperature parameters applicable for analysis are determined by the actual conditions of use.

The type of extraction cell required to conduct each test is determined by time and temperature parameters, the type of food contact material, the potential migrants, and the simulants required. Typical extraction cells used are shown in Figure 1. After extraction, the additives are characterized and quantitated using gravimetric, chromatographic, and spectroscopic techniques. The methods used are very sensitive and can detect indirect additives in the low ppb range.

Testing recycled materials

The FDA requires that recycled materials must meet the existing purity standards for virgin material. However, safety factors for using recycled materials in packaging are more complex. As a result, even though the regulations are generic, additional characterization and testing beyond that used for virgin materials are required when a recycled product is tested to ensure it is thorough and accurate. In 1992, the FDA published its "Points to Consider for the Use of Recycled Plastics in Food Packaging: Chemistry Considerations."[7] In 1995, a task force formed by the National Food Processors Association and Society of the Plastics Industry, Inc., published their "Guidelines for the Safe Use of Recycled Plastics for Food Packaging Applications." These two sets of guidelines are based upon the collective experience of both the FDA and members of the two associations. The guidelines establish some standards from which to develop a testing strategy.

When evaluating the safety of recycled materials, the main concerns are chemical contamination, structural integrity and microbial contamination. These hazards are of more concern with plastics than with glass or metal. This is due to the porous nature of the containers and the fact that plastic containers are often re-used in the home for storing motor oil or for mixing chemicals such as pesticides.

Analytical protocols are being developed to demonstrate that contaminant levels in packaging made from recycled materials are sufficiently low to ensure that they can be used safely. The FDA uses three classifications to delineate the approaches to recycling plastic for packaging.[8] These classifications help determine exact methods of testing required.

- Primary (1°) recycling is the use of industrial scrap and salvage not yet used by consumers.
- Secondary (2°) recycling is physical reprocessing of used materials--washing, vacuum and heat treatment, grinding, melting, and reforming.
- Tertiary (3°) recycling involves chemical reprocessing, purification by a variety of techniques, and repolymerization.

Packaging composed of 1° recycling materials is not considered a hazard to consumers if it is produced according to Good Manufacturing Practices specifications. Manufacturers who produce 2° recycled packaging must have control over the source of the recycled resins and determine restrictions on the types of foods that can be packaged. Packaging made of 3° recycled materials is purified in reprocessing. The high temperatures and solvent baths used in the 2° and 3° processes effectively eliminate exposure to microbiological contaminants.

One of the key components of the FDA's report and the Plastics Recycling Task Force was the identification of a set of relatively non-toxic surrogates (i.e., model compounds) for the estimated 60,000 substances that could potentially be found in recycled plastics. The compounds were chosen based on volatility and solubility parameters. This matches the FDA approach of using volatile vs. non volatile and polar vs. non-polar compounds. This was suggested for two primary reasons:

1. The solubility of one substance in another substance is a function of the relative polarities of the two
2. The volatility of a substance is a function of its vapor pressure

Some of the suggested surrogates and their corresponding classification are:

- Toluene   Volatile, nonpolar
- Chloroform   Volatile, polar
- Lindane   Nonvolatile, nonpolar
- Diazinon   Nonvolatile, polar
- Disodium monomethylarsenate   Toxic organo-metal salt

In addition to these surrogates, successful studies that have been acceptable to the FDA have been conducted using other compounds. For example, benzophenone has been used in place of diazinon as a nonvolatile, polar surrogate and phenyldecane and tetracosane have been used as nonvolatile, nonpolar surrogates. In addition, three alternative compounds have been used as toxic salt surrogates: calcium monomethylarsenate (CMMA), copper (II) ethylhexanoate, and zinc stearate. The use of these alternative surrogates provides several benefits. In most cases the alternates are less toxic thereby increasing safety. In addition, some of the surrogates can be used to simulate more that one type of compound. For example, CMMA may be used as both a nonvolatile-nonpolar surrogate as well as a toxic salt.

Study Phases

Although the specifics of each study may vary, a typical study for recycled polyethylene terephthalate (PET) may consist of four phases. The phases conducted are dependent upon the levels of simulants found.

- Phase I: Flake contamination
- Phase II: Flake analysis and method validation (before and after processing)
- Phase III: Bottle or plaque migration study
- Phase IV: Extract analysis and method validation

If, after processing, a level of less than 217 ppb is found in flake, migration testing (i.e., Phase III) may not be required. This value may be calculated using the following factors provided by the FDA.[6]

- A consumption factor (CF) of 5% for PET
- A PET mass/surface area of $0.46 g/in^2$.
- A 10 g food/in^2 surface area contact

Therefore, 0.5 ppb in the diet calculates to 10 ppb in the food (0.5 ppb/0.05 = 10 ppb) and 10 ppb calculates to a maximum of 217 ppb contamination
(10×10^{-9} g/g $\times$ 10 g/1 in^2 $\times$ 1 in^2/0.46g = 217 $\times 10^{-9}$ g/g = 217 ppb).

Nonvolatile analysis

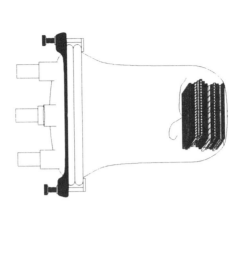

One-sided extractions

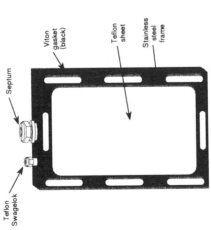

Septum

Teflon Swagelok

Viton gasket (black)

Teflon sheet

Stainless steel frame

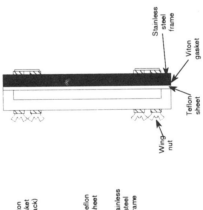

Stainless steel frame

Viton gasket

Teflon sheet

Wing-nut

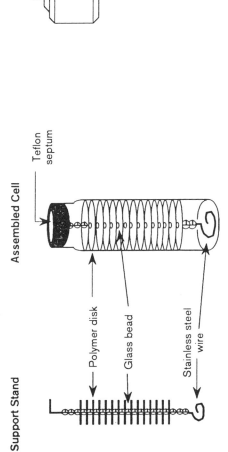

Volatile determinations

Ethanol solutions heated past boiling point

Support Stand

Assembled Cell

Teflon septum

Polymer disk

Glass bead

Stainless steel wire

Figure 1. Types of extraction cells.

The bundles pictured in the "Nonvolatile analysis" and "Ethanol solutions heated past boiling point" diagrams are composed of test material sheets and/or plaques that are separated by stainless steel wire mesh to allow the extraction solution to flow freely around the test material.

Example Studies

Study 1 is an example of data generated from the use of contamination cocktails to evaluate a recycle process. In this study, cocktail compounds of chloroform, toluene, lindane, and CMMA were utilized (Table I).

Table I. Recycle Process Study 1 Contamination

Cocktail Compounds	% Weight	Flake Exposure
Chloroform	49.5	
Toluene	49.5	2 wks @ 40°C
Lindane	0.5	
CMMA	0.5	

The PET resin was immersed in pure cocktail and stored for two-weeks at 40°C with periodic agitation. The cocktail was then drained-off and the resin air-dried. The contaminated resin was sampled at various stages in the recycling and clean-up process. The resin was subsequently formed into sheets and bottles. Results show that in this example, the recycle process did not reduce the chloroform concentration. The toluene and lindane values in the processed resin were somewhat reduced while the CMMA concentration was significantly lowered to 10 ppm (Table II).

Table II. Recycle Process Study 1 Results (ppm)

	Chloroform	Toluene	Lindane	CMMA
Contaminated Resin	6,000	3,000	200	800
Processed Resin	7,000	2,000	100	10

In Study 2 the surrogate compounds were used individually to intentionally contaminate the PET flake. The contamination conditions for this study can be found in Table III.

Table III. Recycle Process Study 2 Contamination

Compounds	Flake Exposure
Chloroform	2 secs @ ambient
Toluene	6 hrs @ ambient
Lindane	2 secs @ ambient
CMMA	2 wks @ 40°C

After exposure, each individual PET resin was drained and air-dried. It was then mixed with virgin PET and processed into tubes, preforms, and bottles. Each of these forms was analyzed at various stages of the process to measure the level of contamination during the clean-up process. As the data in Table IV show, the first step in the process resulted in the greatest reduction in contamination values. Further processing to the preform and bottle stage resulted in no change.

Table IV. Recycle Process Study 2 Results (ppm)

	Chloroform	Toluene	Lindane	CMMA
Flake	5,000	2,500	8,000	11,000
Tubes	2,000	100	300	300
Preforms	1,000	50	100	120
Bottles	1,500	50	150	120

A summary of the analytical methods used to conduct these studies is included in Table V.

Migration Studies

Generally, laminates which have layers composed of recycled material that do not come in contact with the food product or which contain virgin material as a functional barrier, can be used without concern for contamination. However, regulations require that they must be proven to be safe. These products are tested in extraction studies using intentionally contaminated resins similar to those used in standard Phase III and IV tests. After contamination, analyses are conducted to determine the potential migration through the food-contact layer.

Migration studies for PET are conducted using food simulants at 40°C for 30 days. Sampling intervals are 1, 3, 10, and 30 days. Typically, only Day 10 and Day 30 intervals are analyzed. Different detection methods are used for PET or ethanol extraction.

Validation of these studies is accomplished using the following FDA recommendations:

- Standard additions and recovery calculations at 1/2, 1, and 2 times the amount detected.
- If not detected, validation is conducted at the Limit of Detection
- Acceptable recoveries
  <100 ppb   60 - 110%
  >100 ppb   80 - 110%

Continuing Challenges

The study of packaging design and technology is an ever-changing field. New developments must be vigorously investigated to ensure consumer safety. To that end, tests of food packaging materials must be carefully designed and implemented. Researchers are testing newer degradable materials to determine how much they break down and what components, might migrate into food products. Although some have labeled the FDA approach as conservative, most researchers, regulators, and manufacturers agree that the new guidelines are a major step forward in adopting a process to produce safe recycled plastic products thereby helping to alleviate a major environmental concern.

As new methodologies are developed to accomplish this task, we can more accurately document the safety of packaging made from recycled materials. With the knowledge

Table V. Analytical Method Summaries

Lindane

PET	Ethanol	Detection
1g PET with 10 mL of triflouracetic acid (TFA). TFA solution is extracted with hexane and backwashed with water, evaporated and diluted in hexane.	50mL of 10% ethanol is extracted with hexane, evaporated, and dissolved in hexane.	GC using a DB-17 column with electron capture detection

CMMA

PET	Ethanol	Detection
1g PET with water in a Parr bomb and heated to 260°C for 3 hrs.	25 mL of ethanol extracts are pipetted into a vial.	Arsenic determined by hydride generation and atomic absorption

Chloroform

PET	Ethanol	Detection
1g PET with 5 mL of sulfuric acid in headspace vial	Solid phase micro extraction (SPME) fiber is used on ethanol extracts. The fiber is thermally desorbed in the GC injection port.	GC using a DB-wax column with electron capture detection

Toluene

PET	Ethanol	Detection
1g PET with 10 mL methylene chloride is shaken for 24 hours Extract is filtered, passed through a silica gel column into a Kuderna-Danish (KD) flask and concentrated.	Solid phase micro extraction (SPME) fiber is used on ethanol extracts. The fiber is thermally desorbed in the GC injection port.	GC using a DB-1 column with flame ionization detection

Benzophenone

PET	Ethanol	Detection
1g PET with trifluoroacetic and shaken. Solutions are extracted with hexane, backwashed with water, evaporated, and dissolved in methanol.	Aliquots of ethanol solution are injected directly into the HPLC.	HPLC using a C-18 column with ultraviolet detection

Copper (II) Ethylhexanoate

PET	Ethanol	Detection
PET placed in vycor crucible with 4% HCl and heated to 400°C overnight	Aliquots of ethanol solution are injected directly into the graphite furnace	Flame atomic absorption for high levels Graphite furnace atomic absorption for low levels

gained from these studies, the packaging industry can be more efficient in developing additional recyclable products without increasing the potential for migration of harmful indirect food additives.

Literature Cited

1. The Society of the Plastics Industry. *Facts and Figures of the U.S. Plastics Industry.* The Society of the Plastics Industry, Inc., Washington, D.C., 1995, pp. 25-26.
2. Environmental Protection Agency. *Characterization of Municipal Solid Waste in the United States, 1994 Update.* Environmental Protection Agency, Washington, D.C., 1994, pg. 4.
3. Environmental Protection Agency. *Characterization of Municipal Solid Waste in the United States, 1994 Update.* Environmental Protection Agency, Washington, D.C., 1994, pg. 6.
4. The Society of the Plastics Industry. *Facts and Figures of the U.S. Plastics Industry.* The Society of the Plastics Industry, Inc., Washington, D.C., 1995, pg. 91.
5. Bayer, Forest L, "The threshold of regulation and its application to indirect food additive contaminants in recycled plastics." *Food Additives and Contaminants,* 1997, Vol. 14, No. 6-7, pp 661-670.
5. Federal Register, "21 CFR 176.170 Components of paper and paperboard in contact with aqueous and fatty foods." *Code of Federal Regulations 21 Parts 170 – 199,* U.S. Government Printing Office, Washington, D.C., 1998, pg. 201.
7. Food and Drug Administration. *Points to consider for the use of recycled plastics in food packaging: Chemistry considerations.* Center for Food Safety and Applied Nutrition, Washington, D.C., 1992, pp. 1-3.
8. Food and Drug Administration. *Points to consider for the use of recycled plastics in food packaging: Chemistry considerations.* Center for Food Safety and Applied Nutrition, Washington, D.C., 1992, pg. 5.

Chapter 11

New Test Methods for Highly Permeable Materials

Robert L. Demorest, William N. Mayer, and Daniel W. Mayer

MOCON Inc., 7500 Boone Avenue North, Minneapolis, MN 55428

Both developers and end-users of barrier polymeric films have long searched for more precise methods to test the permeability of their materials. Today, modern, repeatable test methods are in daily use for good barrier testing. The same has not been true for highly permeable material applications, until now. Fresh-cut, ready-to-eat salads depend on high oxygen permeable films to maintain their freshness while they continue to respire in the package. Similarly, disposable diapers require high water vapor permeable films to "breathe" while keeping their outer surfaces dry to the touch.

Both of these applications, and many others, are benefitted by new test methods and apparatus which precisely, repeatedly measure the high permeation rates desired by overcoming the measurement shortcomings of the past. This paper outlines these problem areas and presents data using the new test technologies.

To better protect many food products, packaging companies have been striving to create better barriers to moisture, oxygen and other gases. There is, however, a segment of the industry that is interested in highly permeable materials. These companies desire large amounts of gases to be able to pass through their materials.

Fresh cut produce, such as ready-to-eat salads, continues to respire in the package. This demands a material with a relatively high oxygen permeation rate into the package in order to keep the lettuce fresh. Disposable baby diapers require high water vapor permeation rates through the outer surface to permit evaporation and to avoid a clammy outer skin feeling. In both of these cases, unique objectives have created unique challenges. One of the challenges has been the development of standardized tests to accurately measure the high permeation rates.

Oxygen Transmission Rates

For the past 20 years, most oxygen transmission (and permeation) rates have been determined using ASTM[1] test method D-3985. This is an isostatic method that measures the small amount of oxygen which is permeating through a 4" x 4" flat film sample installed in the test apparatus. A sensitive coulometric sensor, employing Faraday's Law, releases four electrons for each O_2 molecule it sees. This current is dropped across a load resistor, and a computer takes over from there. The film is held at a precise temperature and humidity because both of these parameters affect the permeation rate of oxygen through many polymers. One side of the sample is constantly exposed to a slow flow of 100% O_2, and the other side to a slow flow of 100% N_2. The oxygen permeating through the film from one side to the other is swept by the nitrogen carrier gas into the coulometric sensor. This test method was designed for low permeating, (also called high barrier) materials such as those with transmission rates typically in the range from 0.0003 to 50cc/100 $in^2 \cdot$ day $\cdot$ atm (0.005 to 775 cc/m^2 $\cdot$ day $\cdot$ atm). A diagram of a typical test cell is shown in Figure 1.

With time, the amount of oxygen permeating reaches a steady-state, or equilibrium amount, and the equilibrium O_2 permeation rate is established, as depicted in Figure 2. Typical test conditions are 37.8C (100F) and 90% RH, and the units used are cc /100 $in^2 \cdot$ day $\cdot$ atm (or cc/m$^2 \cdot$ day $\cdot$ atm).

As mentioned previously, not all products require a low O_2TR. Some, such as fresh cut salads need very high oxygen transmission rates.

High Oxygen Transmission Rates

In the past, those laboratories that needed to test materials with high O_2 transmission rates had to use special techniques to reduce the measured transmission rate so that it would be within the range of the detector. These included such things as masking the sample to a smaller surface area, or using a lower oxygen concentration as the permeant to reduce the driving force. Although somewhat successful, these methods suffered from poor repeatability and poor correlation between labs.

Over the past three years, new methods and instrumentation have become commercially available to test materials in much higher O_2 transmission ranges. It is now possible to test all the way up to 10,000 cc/100 $in^2 \cdot$ day $\cdot$ atm. A typical range for these high transmission rate materials is 50 to 10,000 cc/100 $in^2 \cdot$ day $\cdot$ atm.

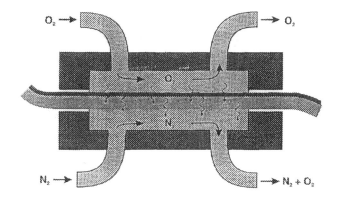

Figure 1: ASTM D-3985 Isostatic Oxygen
Permeation Test

Barrier Sample Coming to Equilibrium

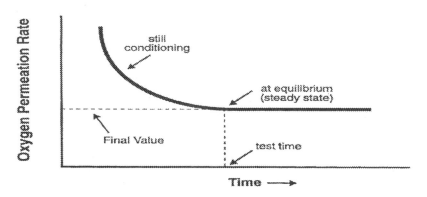

Figure 2: Oxygen permeation rate relative to time.

High permeating, continuous (non-perforated) materials such as those made from metallocene resins, have entered the market place over the past few years. As the films have been evolving, so have the test methods and apparatus. New sensors had to be developed which could measure these high amounts of oxygen. New calibration techniques were needed and new test protocols had to be perfected.

Figure 3 shows the relative respiration rates for a variety of fresh fruits and vegetables. (2) Broccoli has a very high respiration rate when compared to celery sticks and green pepper. It will require a packaging material with much higher transmission rate in order to preserve the freshness. While the package must have a high transmission rate for oxygen, the water vapor transmission rate must be low to maintain the moisture content of the produce.

Using the new technique that has been developed, a film that was designed to package cut celery was tested. The material was a 3.1 mil coextruded film. The results of duplicate tests are shown in Figure 4.

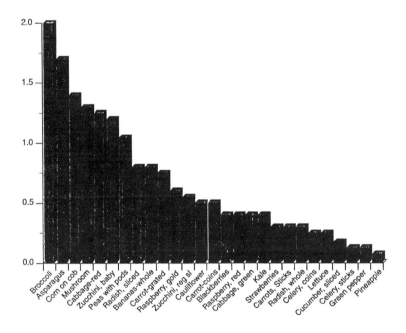

Figure 3: Relative Respiration Rates of Fresh Produce

(Reproduced with permission from reference 5. Copyright 1997.)

As can be seen, the transmission rate at 6C was less than half of that when measured at 23C. This shows the importance of testing at the intended temperature of use. To determine whether or not humidity had an impact on the transmission rate, the test was repeated at 6C using 90%RH instead of dry conditions. There was no difference in the oxygen transmission rate at 6C between dry and humid conditions.

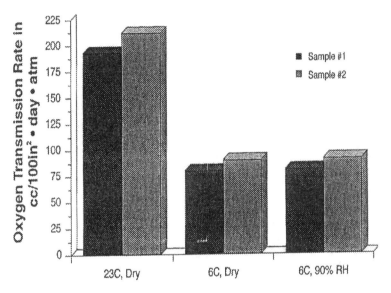

Figure 4: Oxygen Transmission Rate
Through Celery Packaging Film

120

The significance of this is that it is very important to test materials at the conditions of real-world use. This celery film is not sensitive to moisture, but transmits far less oxygen at 6C than it does at room temperature. This information is vital to knowing how much oxygen is entering the package at any point in time. If these films are specified at 23C, but are used in the real-world at 6C, problems can occur which could affect the shelf-life and safety of the product.

Water Vapor Transmission Rates

Let's turn our attention to the measurement of WVTR. In North America, WVTR has been measured since the 1940's with the cup test, ASTM E-96,(3) and since the 1980's with ASTM F-1249.(4) The useful ranges for these tests are 0.65 to 3.23 and 0.002 to 6.45 g/100 in^2 · day for E-96 and F-1249, respectively.

High Water Vapor Transmission Rates

As with oxygen transmission, not everyone is designing a package which requires a good H_2O barrier. Some applications call for very high transmission rates of moisture, which presents a problem for the traditional tests. The moisture transmits through the sample so quickly that it is difficult to maintain the desired gradient from one side of the film to the other. To accurately measure the transmission rate, the test system should have 100% RH as the driving force on one side of the film, as shown in Figure 5.

This cannot be achieved due to the rapid loss of water through the sample, and due to the slow replacement of H_2O molecules from the water reservoir below the sample. The result is an unknown RH at the lower surface of the sample. Additionally, on the top surface of the sample, moisture is gathering and raising the 0% RH to some unknown higher value.

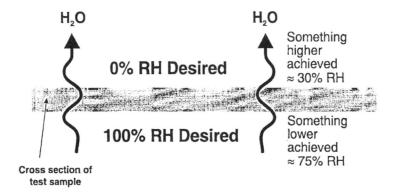

Figure 5: Moisture gradient for water vapor transmission testing

In the example shown in Figure 5, the sample is seeing a gradient of perhaps only 45%RH (the actual value is unknown) instead of the desired 100% gradient. This lower gradient results in dramatically lower WVTR values than the actual. This situation with a lower gradient across the film creates errors in both methods, ASTM E-96 and ASTM F-1249.

It is important to understand the difference between porosity and permeability, and the measurement of each. A porous material has holes in it, and a permeable membrane does not. Of course, a material can exhibit both porosity and permeability simultaneously.

Porosity is the measure of a gas flow (such as water vapor) through a barrier material (such as paper) when a static pressure difference exists across the barrier as shown in Figure 6. This flow can be measured in different ways, and is usually expressed in Gurley seconds or Darcies. This is not a real-world test if the sample has the same static pressure on each side in real use. Also, this test does not measure permeability, diffusion, or transmission rate.

Permeability is the measure of a gas moving through a barrier material when there is equal static pressure on both sides of the barrier, but where the partial pressure of the permeant is different as shown in Figure 7. This is a real-life situation with many non-wovens, textiles, microporous membranes, and papers. This type of test measures the actual permeability, diffusion, and transmission rate of water vapor gas through barrier materials, both porous and non-porous. We are talking about real-world situations when the static pressure is exactly the same on both sides of the material.

In response to these gradient and pressure problems, new methods and apparatus have been developed. New sensors had to be developed, as well as software, hardware, and methodology. A comparison of the new method to ASTM E-96 is shown in Figure 8. The results are for six different materials, designated A - F.

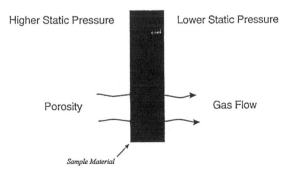

Figure 6: Diagram of conditions to measure porosity

(Reproduced with permission from reference 5. Copyright 1997.)

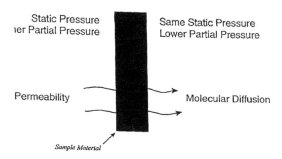

Figure 7: Diagram of conditions to measure permeability
(Reproduced with permission from reference 5. Copyright 1997.)

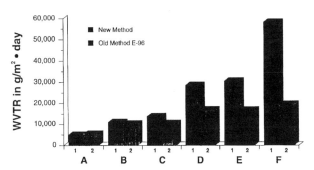

All six materials are different. All samples labeled #1 were tested using the new method, and all labeled #2 were tested using the old E-96 method.

Figure 8: Comparison of E-96 and New High
WVTR Test Method, on Six Materials

(Reproduced with permission from reference 5. Copyright 1997.)

As expected, the difference in the answers obtained by the two methods is small for materials with a relatively low WVTR, but is much larger at the higher end of the range. Near 5,000 $g/m^2 \cdot$ day, both methods essentially agree, adding credibility to the new method. However, as the rates increase, the gradient cannot be maintained, and the test results begin to fall off dramatically with the E-96 cup test for the reasons previously described.

This new method can be used to test any high moisture transmitting material including non-woven and woven materials, perforated films and papers. It is now possible to test over a range from 500 - 100,000 $g/m^2 \cdot$ day (32.3 - 6,452 $g/100 in^2 \cdot$ day). An ASTM standard based on this new technology is under development. The significance of this new data is that, for the first time, repeatable, reliable values for very high WVTR can now be determined.

New test methods and apparatus now exist to measure both high oxygen and water vapor transmission rates through today's modern "breathable" materials. The new precision available should greatly improve development, use, and applications for these exciting new materials.

References:

1.   ASTM is the American Society for Testing and Materials, 1916 Race Street, Philadelphia, PA 19103, USA.

2.   Source unknown.

3.   ASTM E-96 is also known as the gravimetric cup test.

4.   ASTM F-1249 is the isostatic test method employed in the MOCON PERMATRAN-W® product line.

5.   FUTURE-PAK® is a registered trademark of George O. Schroeder Associates, Inc. FUTURE-PAK '97 took place in Chicago, IL USA in October 1997.

6.   PERMATRAN-W® and MOCON® are registered trademarks of MOCON, Inc.

Chapter 12

Mechanism of Aroma Transfer through Edible and Plastic Packagings

Are They Complementary to Solve the Problem of Aroma Transfer?

Jesús-Alberto Quezada-Gallo[1,2], Frédéric Debeaufort[1,3], and Andrée Voilley[1]

[1]ENS.BANA, Laboratoire de Génie des Procédés Alimentaires et Biotechnologiques, Université de Bourgogne, 1 Esplanade Erasme, 21000 Dijon, France
[2]CONACyT, Mexico
[3]I.U.T. Génie Biologique, Boulevard du Dr. Petitjean, BP 510, 21014 Dijon Cedex, France

Aroma compounds have a strong affinity for the most of plastic and resin polymers which contribute to their relative high permeability to flavors. Furthermore, some natural polymers such as polysaccharides and proteins, used as support in the flavoring industry, have high barrier properties against aroma transfer. This work deals with the mechanism of transfer of methylketons, ethyl esters and terpens (aroma molecules usually present in food) through edible films mainly composed of polysaccharides and proteins from vegetal origin. Sorption, diffusion, permeability and structure properties of edible polymers are specially focused and compared to plastic film performances. It appears that the permeability to aroma compounds of edible films depends more on the sorption of the volatile compound and its plasticizing effect than on diffusion, whereas aroma transfer through polyethylene film depends strongly on structural characteristics of both aroma compound and polymer network.

The quality and stability of food products today depends on packaging materials more than ever. Usually, the contact between the food product and the package material does not impact the consumer. However, migration of water, oxygen or of other solutes, such as aromas, can occur through the packaging materials. The latter induces changes in the food quality. Indeed, the loss of aroma compounds could result in a decrease in the flavor intensity or a modification of the aromatic note of the food. Moreover, packaging could modify hedonic characteristics of the food product because of permeation of compounds from the outside or because of the release of low molecular weight compounds from the packaging material which could contaminate the food. These are the reasons why the food industry tends to limit or to control the transfer of small molecules between foodstuffs and the surrounding media by the use of an appropriate package (*1*).

Interactions between Aroma Compounds and Plastic Packaging. The aroma of a food product is comprised of all the volatile compounds that can give an olfactory, and thus a sensory, sensation. These compounds are volatile organic molecules that the human nose can detect even at very low concentrations. A loss or a sorption (ad- and/or absorption) of these compounds by a food product can readily be perceived by the consumer. Therefore, the aroma of a food product must be maintained during storage.

Substances constituting the aromas are molecules which the nature is very similar to those of the main organic solvents used in the food and plastic industries (alcohols, esters, ketones, aldehydes, ...). They have low molecular weight (< 400 g.mol^{-1}) and have a sufficiently high vapor pressure to be partially gaseous in usual storage conditions, and moreover, they are often apolar and thus hydrophobic. Therefore, aromas are able to have a strong affinity for apolar and hydrophobic polymers such as plastics (polyolefines, polystyrene, polyamides, polyvinylchloride, poyesters, ...) (2).

Because of their size, shape and nature, aroma compounds interact with packaging materials. These interactions follow mainly two types of mechanisms :
- adsorption and/or absorption phenomena also called scalping,
- or permeation phenomena such as migration "IN" for which the transfer occurs from outside to inside the packaging, and migration "OUT" corresponding on the loss of solute from the packaged food toward the surrounding medium.

These mechanisms of scalping and permeation often involve a plasticization of the polymer which tends to changes in the appearance (opacification), the loss of adhesion or sealability, partial or total solubilization of the polymer, or again cracks.

In summary, aromas could induce a loss of the food packaging integrity (3,4). In the case of the food product, the latter causes an adulteration of the sensorial quality and a decrease of the shelf life mainly due to microbial contamination, color and texture changes, chemical and enzymatic reactions and specifically losses of flavor and taste or the development of undesirable odor and taste.

Consequences of Interactions on Food Quality. Interactions between packaging and food aroma involved can have an impact on food quality. Indeed, several publications indicated significant loss of the volatile compounds from orange juice stored at 4°C and packed in different types of materials such as high and low density polyethylene (HDPE, LDPE), ethylene vinyl alcohol (EVOH) or polyethylene terephtalate (PET) (5-7). The quantity of aroma loss are between 1 and 9 % for alcohols, between 0.3 and 64% for aldehydes and between 2 and 85% for esters (Table I). These losses are much more important in the case of polyethylene which is not a material recognized for its barrier properties compared to EVOH or PVDC (polyvinylidene chloride) The latter have oxygen permeabilities 1000 or 100 000 times lower than those of polyethylene. However, high barrier polymers are often associated with polyethylene in multilayer complexes because of the thermal sealability of the polyethylene. These multilayered packaging materials, in spite of their low permeability, they have sorbed high quantities of aroma molecules because of their great affinity for the scaleable layer which is always in contact with the food.

Table I: Aroma Compound Losses from Orange Juice because of Scalping in Packaging Based on LDPE, HDPE, EVOH, PET.

Chemical functions of aroma compounds	Percentage of aroma lost after 20 days of storage at 4°C
Alcohols	1 - 9 %
Aldehydes	0.3 - 64 %
Esters	2 - 85 %

Source: Adapted from ref. *5-7*.

Otherwise, sensorial quality of orange juice is decrease of 50% after only one week of storage at 25°C in a paper-aluminum-polyethylene multilayer package (brick). In this case the polyethylene is directly in contact with the juice. After six weeks, the hedonic response is only 20% of the initial value and the orange juice has also lost 25% of its ascorbic acid (Vitamin C). In a publication on the opposite, Stöllman (*8*) showed that the sensorial quality of beverage can be altered by the release of volatile aromatic compounds from PET bottles which contained cola soda or citrus juice prior to be reused. Indeed, a performing washing does not remove all compounds sorbed in the plastic bottles, and thus, the latter cannot be reused for food applications.

Edible Films and Coatings : a Solution ? Most of the works dealing with packaging-volatile compound interactions concern mainly beverages, the same problems occur for other products whatever their state, liquid, viscous or solid.

Consequently, one of the solutions viewed for limiting interactions between volatile compounds and plastics, is to retain aromas inside the food product by adding a supplementary barrier. This can be an edible package, that is to say a protective film, coating or thin layer, having good selectivity against transfers, which is an integral part of the food and which can be consumed with it. This is a macro-encapsulation of the food with the aim to control aroma release (Figure 1).

These edible packaging materials are mainly composed of a film forming substance, which provides cohesiveness, and/or a barrier compound which provides impermeability to the packaging. These substances are usually food ingredients and additives, that is to say carbohydrates, proteins or lipids, alone or in mixtures. The same compound can possess both barrier and film forming properties. An example is gluten which will make edible films with acceptable mechanical properties and high oxygen barrier efficiency.

Functional properties of edible films and coatings strongly depends on their composition. Indeed, protein-based and polysaccharide-based packaging have good organoleptic, mechanical and non condensable gases (O_2, CO_2, N_2) and aroma barrier properties. However, protein- and carbohydrate-based edible packaging are not good at preventing moisture migration. Lipid-based edible films and coatings usually provide a good barrier to moisture.

then injected and analyzed by chromatography (21,22). The main interest of this technique, in the case of dilute solutions, is that it reproduced as closely as possible what occurs between the packaged food product and the surrounding medium. But, because of the limited volume of the analyzed compartment, the concentration gradient applied between the two surface of the film decreases with time, and thus the permeation rate also decreases. In these conditions it is impossible to know accurately the transfer coefficient (transfer rate, permeance or permeability) of the polymer to the volatile compound at the stationary state of transfer. This method is less used than the dynamic one.

The measurement principle remain the same in the case of dynamic methods. However, inner and outer compartments of the permeation cell are continuously swept by a gas (nitrogen, helium hydrogen or air). The inner compartment contains one or several volatile compounds and eventually water vapor, whereas the outer one contains only the dry or wet carrier gas previously cited. The gas flow containing permeant is obtained by bubbling gas through a sample of pure compound or through a dilute solution. The interest in bubbling a gas through pure aroma is the allows you to create a lower concentration and/or to wet the saturated vapor phase by mixing several gas flows before the inner compartment of the permeation cell. Figure 3 presents the apparatus developed by Debeaufort and Voilley (14). This system allows the simultaneous determination of the permeabilities of several aroma compounds as well as the permeation of water vapor and to non condensable gases when the latter is used as the bubbling gas.

The main advantage of the dynamic method is that the aroma concentration gradient remains constant and it is possible to determine the « true » permeability of a film at steady state conditions. Nevertheless, a main precaution has to be taken in this method. It is necessary to obtain aerodynamic conditions which guarantee that the effect of stagnant layers is negligible as shown by Debeaufort and Voilley (14). But, because of the high Reynolds value to prevent a stagnant layer, the permeability obtained in these conditions does not represent the reality of the food interacting with its package.

The dynamic isobaric method is the one which has been developed to be marketed. The MAS2000 (MAS Technologies, Zumbrota, MN, USA) simply uses an FID detector at the exit of the outer compartment. This gives a signal directly proportional to the aroma transfer rate, but it only permits you to work with the saturated aroma vapor phase in the inner compartment. If a mixture of aromas is used, it gives only a global value of the permeability because there is no column to separate the compounds before the detector. The methodology used by the Lyssy GPM 500 (LYSSY, Zollikon, Switzerland) is based on the same principle as the one given in Figure 3. It does not work in moist conditions. Several adaptations of these techniques were proposed as a function of the desired objectives. Indeed, (23) placed a cryofocuser between the outer compartment of the cell and the injection port of the gas chromatograph. This allows them to concentrate permeated aroma vapors before analyses, and thus allows them to obtain a very efficient detection level. It is this system that MOCON (Mocon Modern Controls Inc., Minneapolis, MN, USA) used in the Aromatran apparatus.

This technique is the most used for the determination of aroma permeation through plastic packaging (15, 24-28), but also for edible films and coatings (12,29).

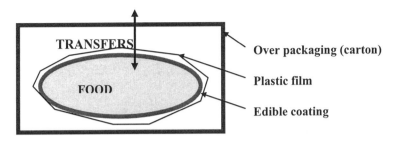

Figure 1: Complementarity between edible coating and traditional packaging against the transfers of volatile compounds.

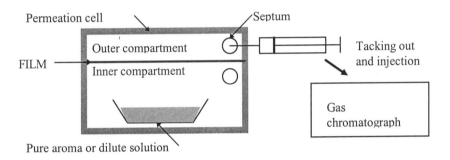

Figure 2: Isobaric static method for the determination of aroma permeability of films.

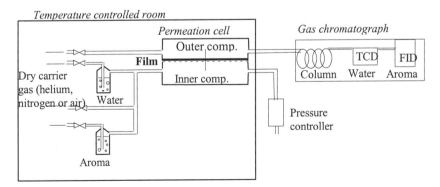

Figure 3: Dynamic isobaric system for the determination of volatile transfer rate by gas chromatography analysis.

Table II: Sorption (S), Diffusion (D) and Permeation (P) Coefficients of Aroma Compounds in some Edible and Plastic Films at 25°C.

Packaging		Aroma compounds	S (g.m^{-3}.Pa^{-1})	D (10^{-13} m^2.s^{-1})	P (10^{-6} μg.m^{-1}.s^{-1}.Pa^{-1})
edible	Methylcellulose	1-octen-3-ol	227	4.25	122
		2-pentanone	15	4.0	19
		2-heptanone	140	2.1	39
		2-octanone	600	0.7 - 4.2	338
		2-nonanone	1139	1.2 - 8.7	420
		ethyl acetate	2.22	2.07	128
		ethyl butyrate	4.86	0.43	119
		ethyl isobutyrate	2.42	0.39	106
		ethyl hexanoate	100.9	0.19	668
		d-limonene	10	-	10
	Wheat gluten	1-octen-3-ol	64	1.35	4.6
		2-pentanone	7	2.1	0.12
		2-heptanone	24	0.3	0.50
		2-octanone	42	-	< 0.005
		ethyl acetate	1.06	2.19	0.059
		ethyl butyrate	1.30	2.73	0.670
		ethyl isobutyrate	0.88	1.87	0.04
		ethyl hexanoate	12.5	-	< 0.005
	whey proteins	d-limonene	-	-	0.0003
Cellulosic	Cellophane	1-octen-3-ol	4	2.7	0.03
plastic	LDPE	1-octen-3-ol	75	1.7	51
		citronellol	829	2.6	-
		menthol	325	1.2	-
		2-pentanone	6	2.8	9
		2-heptanone	24	4.5	239
		2-octanone	129	0.2	133
		2-nonanone	334	0.2	165
		d-limonene	375	4.3	106
	HDPE	citronellol	908	0.05	-
		menthol	612	0.06	-
		d-limonene	256	0.5 - 0.9	16
	BiO-PP	citronellol	572	0.013	0.15 - 1.2
		menthol	276	0.007	-
		d-limonene	523	0.03	-
	PET	d-limonene	-	0.0006	0.000015
	EVA	d-limonene	-	-	194
	Co-VDC	d-limonene	-	-	0.18
	EVOH	d-limonene	-	-	0.00001

Source: Adapted from ref. 9-17.

The barrier efficiency of an edible packaging against aroma transfer is good since it has very few affinity for the compound and since it decreases their transfer speed. that is they have very low solubility, diffusion and thus permeability coefficients. From the results obtained in the laboratory and from literature, S, D, and P values of edible films to D-limonene (lemon odor), to 1-octen-3-ol (mushroom), 2-pentanone (fruity), to 2-heptanone (banana), to 2-octanone (herbaceous, cheese) and to 2-nonanone (rose) are of the same order of magnitude as that observed for plastic films (Table 2). Table 3 summarizes the order of magnitude of solubility, diffusion and permeability coefficients in edible an plastic films ,whatever the chemical nature of aroma compounds.

Table III: Sorption (S), diffusion (D) and Permeation (P) Coefficients of Edible and Polyethylene Films to Volatile Aroma Compounds

Films	Polyethylene	Carbohydrate and protein based edible films
S $(10^6 \ \mu g.m^{-3}.Pa^{-1})$	6 to 830	15 to 1140
D $(10^{-13} \ m^2.s^{-1})$	0.2 to 4.5	0.7 to 8.7
P $(10^{-6} \ \mu g.m^{-1}.s^{-1}.Pa^{-1})$	9 to 240	4 to 420

These works also display that the main factor affecting the barrier efficiency of edible films is the solubility coefficient, *i.e.* the affinity of the aroma compound for the natural polymer whereas, in the case of polyethylene films, diffusion plays an important role in the transfers (*17-19*). Some trials on model food products (model solution of a liquor for stuffed chocolate toffees) showed that a coating composed of whey proteins, wheat gluten and lipids will retain from 60 to 99.9% of the aroma compounds normally lost without edible barriers (*20*).

Measurement of Aroma Barrier Properties of Edible Packaging

The increasing need for efficient barrier packaging requires us to better understand their behavior regarding permeation of oxygen, carbon dioxide, water and organic volatile compounds. Techniques for the measurement of the oxygen and water permeabilities are well known and standardized. Various techniques have been developed for aromas, but the more frequently cited and used are static and dynamic isobaric methods. Isobaric methods consist in placing a film sample between two compartments having the same total pressure but having different partial vapor pressures or different volatile compound concentrations. In all of these techniques, qualitative and quantitative analysis is usually done by gas chromatography.

In the static method, one of the compartments of the measurement cell contains the aroma compound either pure or in solution (Figure 2). The other compartment is airtight, and at regular time intervals, the vapor phase is taken using a gas syringe and

Factors Affecting the Aroma Permeability of Edible Films.

The aroma barrier properties of edible and plastic films depends on many factors. These can be classified in 3 groups :

- Those depending on the characteristics of the volatile compounds such as the chemical nature, carbon number, tridimensional conformation, polarity, solubility, saturated vapor pressure.
- Those related to the polymer such as the nature of the monomer, density, structure, crystallinity, thickness, surface hydrophobicity, etc.
- Those depending on external conditions, *ie* aroma concentration, temperature, pressure, aerodynamic conditions, humidity, presence of other permeants, but also food characteristics such as pH, viscosity, texture, etc.

Factors Related to the Polymer. The nature of the polymer strongly affects its barrier properties (Figure 4). Indeed, the D-limonene permeabilities of plastic films varies 10 000,000 times from EVOH to EVA. This is not related to their hydrophobicity or water sensitivity because both are hydrophilic polymers. In the field of edible packaging, very few data concerning aroma permeability are available in the scientific literature. From Table 3, we can note that protein based edible films seem to be much more efficient than polysaccharide based films.

The permeability is defined as the transfer rate normalized by the film thickness (x) and difference of partial vapor pressure (Δp) of the volatile compound at the surfaces of the film, as described by the following equation :

$$P = TR\frac{x}{\Delta p} = \frac{\Delta m}{A.\Delta t}.\frac{x}{\Delta p} \qquad (1)$$

where Δm, is the amount of aroma compound that permeates during the Δt time across an exposed area A. From this equation, permeability is a constant coefficient and thus the transfer rate, TR, is inversely proportional to the thickness of the material. This relationship is obeyed for most of the plastic polymers for water and non condensable gases. However, plastics and edible polymers interact with aroma compounds, and this relationship often doesn't apply (Figure 5). Indeed, water vapor, oxygen or 1-octen-3-ol transfer rates through a methylcellulose based edible film decrease exponentially when thickness increases up to 50 µm, and for higher thickness, it decreases linearly. This was also observed in the case of low density polyethylene and some aromas (*11*).

Addition of plasticizers within polymeric materials usually provide interesting mechanical properties such as deformability, plasticity, flexibility and pliability. Nevertheless, plasticization involves a swelling of the polymer network, a decrease in the crystallinity and density and thus an increase of the chain mobility which rises also the aroma diffusion coefficient and results in a higher permeability. Figure 6 show a small increase in water (because water is a better plasticizer than polyethylene glycol) and tremendous increases of oxygen and 1-octen-3-ol transfer rates with increasing polyethylene glycol content (used as plasticizer) in methylcellulose-based edible films. This is explained by the high solubility of oxygen and aroma in the polyethylene glycol.

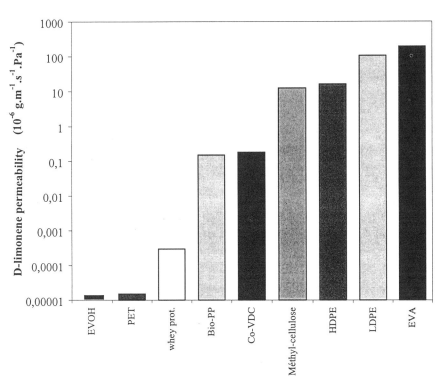

Figure 4: Compared permeability of some plastic and edible polymers to D-limonene at 25°C.

134

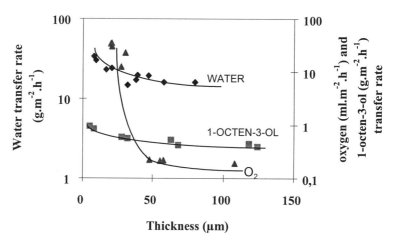

Figure 5: water vapor, oxygen and 1-octen-3ol transfer rates versus methylcellulose film thickness.

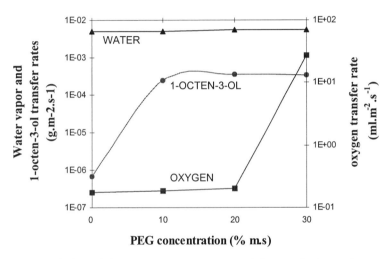

Figure 6: Evolution of the water vapor, oxygen and 1-octen-3-ol transfer rates as a function of the plasticizer content (polyethylene glycol 400) of a methylcellulose based edible films (25°C, 25μm).

Factors Related to the Permeant From Table II, polymer permeability to aroma compounds having the same chemical nature does not always decrease when chain length increases. This is the case with 2-methyl ketones and ethyl esters permeation through protein-based films. Moreover, molecules having the same carbon number (1-octen-3-ol, ethyl hexanoate and 2-nonanone) have very different behavior during permeation. Indeed, permeability to these eight carbon number molecules varies 3 times for LDPE film, 6 times for the methylcellulose film and 1000 times for the wheat gluten film. Moreover, branched molecules diffuse slower than linear ones as observed for ethyl butyrate and ethyl isobutyrate (Table II).

Compounds with the same chemical nature, such as 2-methyl ketones, have very different behaviors as a function of their concentration in the vapor phase in the inner compartment of the permeation cell. 2-nonanone and 2-octanone transfer rate through a methylcellulose film increases linearly with concentration whereas transfer rate of 2 heptanone and 2 pentanone show respectively a sigmoid and logarithmic curves (Figure 7). Indeed, the nature of the interactions between the methyl-ketones and the polymers are not the same when the carbon number change, even for a homologous series of compounds.

Moreover, the same molecule, such as 2-heptanone, reacts very differently as a function of the nature and structure of the polymer network. The permeability of the methylcellulose film is multiplied 10 times when the concentration reaches $15\mu.ml^{-1}$, whereas it is divided by 2 in the case of the low density polyethylene (Figure 8). This ketone has a plasticizing effect on the methylcellulose network while it tends to antiplaticize the polyethylene polymer. This was confirmed by a study of the polymer network by DSC and spectroscopy measurements.

Influence of External Factors. It is well known that an increase in the temperature increases transfers following the Arrhenius law, whatever the permeant and the polymer but only as long as the structure of the latter is not changed by the temperature due to phase transition, partial melting or some other transition.

Conversely, the effect of a co-permeant is less predictable because it can have a synergistic or antagonistic effect on the transfers. So, it is well known that moisture involves a strong increase in the gas permeability (oxygen and carbon dioxide) of synthetic hydrophilic polymers (EVA, EVOH) and edible ones (*30-32*). On the contrary, the effect of humidity on the aroma transfer depends on the nature of both the polymer and aroma compound. If the permeability of methylcellulose or polyethylene films to 1-octen-3-ol increases with the relative humidity (Figure 9), permeability to methyl ketones remains the same and in some cases will decrease (*17*). Moreover, the presence of a compound can have a synergistic effect on the transfer of another compound. Indeed, presence of D-limonene increases significantly oxygen permeability of polyethylene (*3*), as observed for 2-heptanone which makes it easier for the water vapor to transfer through edible films.

Conclusion

Edible packaging seems to be very interesting for the retention of aroma compounds contained in foods, or to prevent migration of off-odors from the

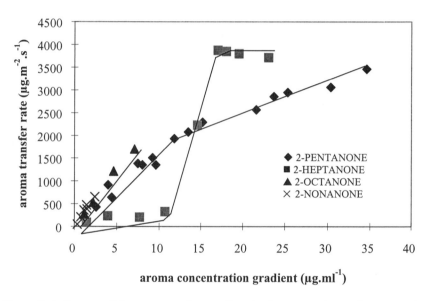

Figure 7: influence of the concentration gradient in the vapor phase of some methyl ketons on their transfer through a methylcellulose based edible film (25°C, 25µm).

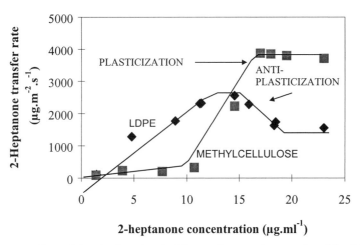

Figure 8: effect of the physico-chemical interactions on the transfer of the 2-heptanone through methylcellulose and low density polyethylene (25°C, 25µm).

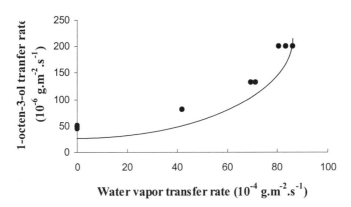

Figure 9: influence of the water transfer rate on the 1-octen-3-ol transfer rate through a methylcellulose edible film (25°C, 25µm).

surrounding medium. However, because of their nature, edible films and coatings always require the use of traditional packaging and overpackaging from petrol or paper origins. Therefore, edible packaging represents an interesting complimentary component for the control of problems found in the field of aroma and plastic packaging. However, many factors affect the barrier properties of edible films and coatings, as noted for some plastic films. For this, a better knowledge of the aroma transfer mechanisms is necessary, to understand the interactions between the food product, edible barrier and plastic package to best maintain the quality of the product.

Literature Cited

1. Mohney, S.M.; Hernandez, R.; Giacin, J.R.; Harte, B.R.; Miltz, J. *J. Food Sci.*, **1988**, *53*, 253-257.
2. Leufven, A.; Stöllman, U. *Z. Lebensm., Unters. Forsch.*, **1992**, *194*, 355-359.
3. Sadler, G.D.; Braddock, R.J. *J. Food Sci*, **1990**, *55*, 587-588.
4. Shimoda, M.; Matsui, T.; Osajima, Y. *Nippon Shokukin Kogyo Gakkaishi*, **1987**, *34*, 535-539.
5. Konczal, J.B.; Harte, B.R.; Hoojjat, P.; Giacin, J.R. *J. Food Sci.*, **1992**, *57*, 967-972.
6. Charara, Z.N.; Williams, J.W.; Schmidt, R.H.; Marshall, M.R. *J. Food Sci.*, **1992**, *57*, 963-967.
7. Dürr, P.; Schobinger, U.; Waldrogel, R. *Lebensmittel-Verpackung*, **1981**, *20*, 91-97.
8. Stöllman, U. In *Food And Packaging Materials - Chemical Interactions;* P. Ackermann, M. Jagerstad, T. Ohlsson Eds., The Royal Society: Cambridge, **1995**, p. 69-73.
9. Becker, V.K.; Koszinowski, J.; Piringer, O. *Deutsche Lebensmittel Rundschau*, **1983**, *79*, 257-266.
10. Franz, R. *Pack. Technol. Sci.*, **1993**, *6*, 91-102.
11. Debeaufort, F. *Etude des transferts de matière au travers de films d'emballages : perméation de l'eau et de substances d'arôme en relation avec les propriétés physico-chimiques des films comestibles*; Ph.D. thesis, ENS.BANA, Université de Bourgogne, Dijon, 1994.
12. Debeaufort, F.; Voilley, A. *J. Agric. Food. Chem.*,**1994**, *42*, 2871-2876.
13. Kobayashi, M.; Kanno, T.; Hanada, K.; Osanai, S.I. *J. Food Sci.*, **1995**, *60*, 205-209.
14. Debeaufort, F.; Voilley, A. *Cellulose*, **1995**, *2*, 1-10.
15. Paik, J.S.; Writer, M.S. *J. Agric. Food Chem.*, 1995, *43*, 175-178.
16. Rouhier, R. *Influence de la sorption de composés d'arômes dans les emballages comestibles sur leur efficacité barrière.* Mémoire d'ingénieur, ENS.BANA, Université de Bourgogne, Dijon, 1997.
17. Quezada-Gallo, J.A.; Debeaufort, F.; Voilley, A. *J. Agric. Food Chem.,*1998. In press.
18. Debeaufort, F.; Voilley, A., Meares, P. *J. Membrane Sci.*, **1994**, *91*, 125-133.
19. Debeaufort, F.; Tesson, N.; Voilley A. In *Food And Packaging Materials - Chemical Interactions*; P. Ackermann, M. Jagerstad, T. Ohlsson, Eds., The Royal Society: Cambridge, 1995, p. 169-175.
20. Gavara, R.; Hernandez, R.J. *J. Plastic Film Sheeting*, **1993**, *9*, 126-138.
21. Apostolopoulos, D.; Winters, N. *Packaging Technol. Sci.*, **1991**, *4*, 131-138.

22. Durand, D. *I.A.A.*, **1996**, *avril*, 211-215.

23. Benet, S.; Ducruet, V.; Feigenbaum, A. *Analusis*, **1992**, *20*, 391-396.

24. Gilbert, S.G.; Hatzimidriu, E.; Lai, E.; Passy, N. *Instrumental analysis of food*, vol 1, Academic Press Inc., 1983, pp 405-411.

25. Delassus, P.T.; Strandburg, G.; Howell, B.A. *T.A.P.P.I. J.*, **1988**, *13*, 177-181.

26. Sensidoni, A.; Peressini, D.; Callegarin, F.; Debeaufort, F.; Voilley, A. *Int. J. Agric. Food. Sci.,* **1998**, in press.

27. Thompson, L.J.; Deniston, D.J.; Hoyer, C.W. *Food Technol.*, **1994**, *48*, 90-94.

28. Doyon, G.; Poulet, C.; Chalifoux, L.; Cloutier, M.; Pascat, B.; Loriot, C.; Camus, P. *I.A.A.*, **1996,** *avril*, 216-220.

29. Miller, K.S.; Krochta, J.M. *Trends in Foods Sci. Technol.*, **1997**, *8*, 228-237.

30. Chomon, P. *L'emballage souple dans l'agro-alimentaire*; Emballage Magazine; Groupe Usine Nouvelle: Paris, 1992; pp 1-120

31. Gontard, N.; Duchez, C.; Cuq, J.L.; Guilbert, S. *Int. J. Food Sci. Technol.*, **1994**, 29, 39-50.

32. McHugh, T.H.; Krochta, J.M. *J. Agric. Food Chem.*, **1994**, *42*, 841-845.

Chapter 13

Prediction versus Equilibrium Testing for Permeation of Organic Volatiles through Packaging Materials

Sara J. Risch[1], William N. Mayer[2], and Daniel W. Mayer[2]

[1]Science By Design, 505 North Lake Shore Drive, #3209, Chicago, IL 60611
[2]MOCON, 7500 Boone Avenue North, Minneapolis, MN 55428

The permeation of volatile organic compounds through packaging materials can be measured by introducing the compound onto one side of the package and measuring the amount that comes through to the other side. If the diffusion rate does not change during the time it takes to reach steady state, there are equations that can be used to predict what the steady state rate will be based on the data collected during the first few hours of testing. Results have shown that many materials may initially appear to follow Fick's first law of diffusion, allowing predictions to be made, however, with time the material will start to change and the actual transmission rate and permeation coefficient will be much higher than predicted. The change in diffusion rate is caused by interactions that occur between the volatile material and the packaging material.

Food packaging materials are designed to protect a product during distribution and storage. The type of packaging material chosen for a specific application will depend on the product characteristics and the desired shelf life. These materials may be specified by a number of different properties, which will indicate the ability of the material to provide the type of protection desired. While there are many different physical properties of a material, the most common specifications for packaging materials that will insure the quality of a food product are water vapor transmission rate (WVTR) and oxygen transmission rate (OTR). In some specific cases, such as carbonated beverages, the transmission rate of CO_2 through the bottle or container is critical to maintaining the quality of the beverage. There are standardized tests for these properties, including several ASTM standards such as D1434 for gas permeability of plastic film, D3985 (2) for oxygen transmission rate, and F1249(3) for water vapor transmission, which are widely accepted.

In recent years, it has been recognized that flavors can interact with packaging materials, potentially resulting in a change in the product over time. While there are no widely accepted specifications for packaging materials based on the flavor transmission rate, the area is one of interest to many food and packaging companies. It is possible to maintain the moisture content of a product and preserve its texture with a package that provides a moisture barrier as well as to preserve other quality attributes with an appropriate oxygen and/or CO_2 barrier. For many products, the next product hurdle that can cause the end of the shelf life is a change in the desired flavor profile. This change can be either a decrease in flavor intensity or a change in the flavor profile.

One of the first instances where products flavor change was noticed and attributed to packaging material was the use of multi-layer brick-packs for orange juice. Based on these reports, there were a number of studies that were carried out to investigate the absorption of orange juice flavor into packaging materials. Hirose et al (4) studied the sorption of the major component of orange oil, d-limonene, by various sealant films. They found that the compound impacted the properties of the sealant layers and that the amount of change was different with different polymers. Another study found that when orange juice was stored in a multi-layer aseptic package at 26 C and 21 C, there were changes in the sensory character detected after one and two weeks, respectively (5). Other studies also looked into the absorption of flavor components by packaging materials (6 – 9). These studies looked at components of orange juice as well as other flavor compounds.

There are two other main interactions between flavor compounds or other volatile organic compounds and packaging materials, which can occur. One of these is permeation through the material. The permeation is dependent not only on the solubility of the organic in the packaging material but also on the diffusivity of that compound through the material. The permeation rate (P) is defined by the equation $P = D \times S$, where is P is expressed in terms of cm^3-mil/100 in^2-day, D is diffusivity (cm^2/sec) and S is the solubility (mg/ cm^3). It should be noted that this equation only holds true when the material being tested follows Fick's first law of diffusion that indicates that the diffusion coefficient does not change with concentration of the permeant. When permeation occurs, desirable components of a flavor can be lost from a product or undesirable compounds may permeate from outside the package into it, resulting in contamination of the product with an off-odor or off-flavor.

The other interaction is migration of components of the package itself into the food product. The compounds that are likely to migrate include residual monomers, low molecular weight additives such as plasticizers, and solvents from either printing inks or adhesives that may have been used. Migration of packaging materials has been studied extensively, particularly with regard to the safety of packaging materials and the products packed in them. One review was written in 1988 (10) that covered much of the literature regarding migration from packaging materials. There are several chapters in this book that address new developments in the testing and prediction of migration from packaging materials for both regulatory and quality issues. Some of the tests today are focusing on the potential for migration from recycled packaging materials.

A number of manual methods have been used to determine transmission rate of volatile organic materials through polymeric films. Hernandez et al (11) reviewed many of the methods that have been tried. Some of these methods can also be used to measure solubility and be used to calculate diffusivity. In one method, one or more volatile organic materials are introduced in the vapor phase to one side of the film in an enclosed test cell (Piringer, O., personal communication). A slow stream of inert gas is passed over the over side of the film and into an organic trap such as Tenax. Periodically, the Tenax trap is removed from the stream and the volatiles extracted from it. These can then be quantified by gas chromatography. Another method that was used to determine solubility involved placing disks of the polymer in a solution of organic materials (12). The concentration of organics in the solution was monitored over time to determine the amount absorbed by the polymer. With these methods, the data was gathered over a series of weeks. When the system reached steady state, that is when either a constant amount of volatiles were passing through the film or when no more was absorbed by the polymer disks in solution, the transmission rate and solubility could be determined.

While these methods are reliable, they are also time consuming and labor intensive. As with many other analytical techniques, there was a desire to automate the test procedure and develop a method to predict P, S and D instead of waiting for the test to come to steady state. There are equations that can be used to predict the steady state values based on data collected at regular intervals during the initial stages of testing. One of these is the half-time method (13). The equations depend on the material being Fickian, that is that the diffusion rate does not change with concentration of the permeant. There is an approved ASTM method for measuring volatile organic compound permeation through packaging materials (14). With organic vapor permeation through packaging materials, the time to reach steady state can be months or even years. Many people want to use the prediction mode of the method described by Pasternak et al (13) to obtain results in much shorter period of time than waiting for steady state permeation. This study was undertaken to determine the validity of using a prediction method versus taking the test to steady state.

Materials and Methods

One commonly used packaging film was selected for testing. This material was oriented polypropylene (OPP), supplied by QPF, Streamwood, IL. Two volatile organic compounds were tested. One was methyl ethyl ketone (MEK) which is a solvent commonly used in printing inks and the other was d-limonene, a major component of citrus oils. The testing was accomplished by placing a sample of the film into a test cell in an Aromatran 1A (Modern Controls, Minneapolis, MN). The permeant was introduced to one side of the film at a constant concentration. The stream of MEK was produced using a cylinder certified to contain 1000 parts per million (ppm) MEK on a volume/volume basis. For the d-limonene, holding a pure sample of the desired permeant at a specific temperature and sparging with a stream of nitrogen generated the stream of permeant at a constant concentration. The other side of the film was constantly swept with a stream of helium that went directly into a flame ionization detector. The detector had been calibrated using a known concentration of permeant. Readings were taken on a regular basis.

Results and Discussion

The transmission rate for MEK is shown in Figure 1. As the testing time increased, the transmission rate also increased until steady state had been reached. As can be seen, the results follow the typical S-shaped curve that is produced by a material that follows Fickian behavior. The transmission rate of MEK as steady state is 153 ul/100 in^2/day. If the transmission rate was predicted at 90 minutes using Pasternack's equation (13), it would have predicted the correct steady state results. In contrast, the predicted results for d-limonene at a concentration of 2473 ppm show that the final, steady state transmission rate if the results were to be predicted after two hours would be approximately 1.3 x 10^6 ug/M^2/day (Figure 2). As the test continued to run, the d-limonene interacted with the film and the transmission continued to increase, not reaching the steady state that would have been predicted. The results of the test, run for a total of 95 hours is shown in Figure 3. The actual value is 3.6 x 10^6 ug/M^2/day.

Similar test were run at two lower concentrations, each one order of magnitude lower than the previous test. Similar results were observed. At first, it appears that the material follows Fickian behavior so that the transmission rate can be predicted after less than two hours of testing. Instead of reaching the predicted steady state, the d-limonene shows evidence of interaction with the film and the transmission continues to increase up to a level that is approximately three times higher than what would have been predicted. The results for the prediction and steady state are shown in Figure 4 for d-limonene at 253 ppm and in Figure 5 for d-limonene at 45 ppm. On each graph, the predicted value is shown after 1.5 hours of testing and the curve is continued to show the steady state results. As can be seen, even at the lowest concentration tested, the material does not follow the behavior that would be predicted. The actual steady state value at all three concentrations is from 1.5 to 3 times higher than what would have been predicted had the test only been run for one and one-half hours.

Further testing was conducted at 30 C using d-limonene at 460 ppm to determine if the results were temperature related. The results of this test indicated that predicted value would be fairly close to the steady state value. It should be noted that the apparent steady state value did start to drift from what appeared to be the steady state value after about 3.5 days of testing.

The work presented here is a preliminary study to investigate the use of prediction testing as compared to taking the test of permeation to steady state. The results at the higher temperature indicate that the predicted value will be lower than the actual value. It is possible that the higher temperature results in changes in the film that in turn affects the permeability. It is important to understand the potential for interaction between the polymer and permeant when doing any testing to determine permeability, solubility and diffusivity. The research emphasizes the fact that unless the interaction between a given permeant and polymer are fully understood, using a prediction method can give erroneous

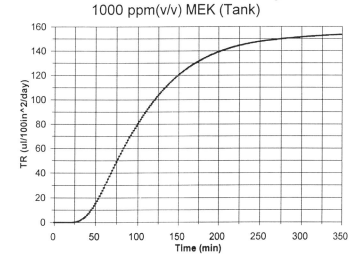

Figure 1. Transmission rate curve for MEK through OPP at 25 C, permeant concentration at 1000 ppm (vol./vol.).

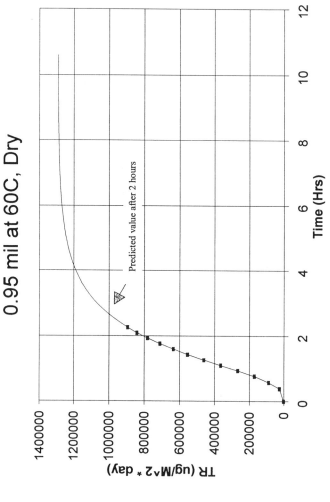

Figure 2. Transmission rate curve for d-limonene through OPP at 60 C after two hours testing to show predicted value of steady state, permeant concentration at 2473 ppm (vol./vol.).

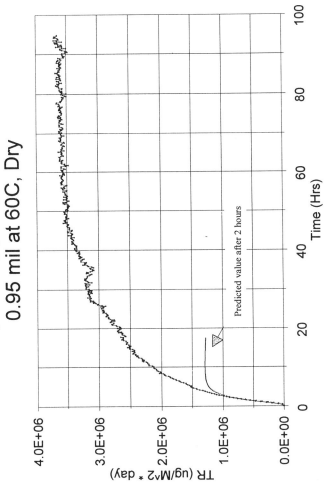

Figure 3. Transmission rate curve for d-limonene through OPP at 60 C showing steady state results, permeant concentration at 2473 ppm (vol./vol.).

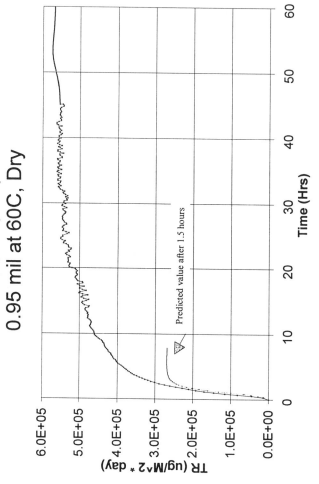

Figure 4. Transmission rate curve for d-limonene through OPP at 60 C showing predicted and steady state results, permeant concentration at 253 ppm (vol./vol.).

149

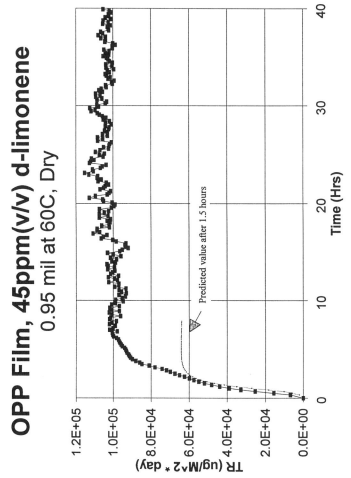

Figure 5. Transmission rate curve for d-limonene through OPP at 60 C showing predicted and steady state results, permeant concentration at 45 ppm (vol./vol.).

150

results. The desire is to rapidly test materials but this does not always yield the correct information. It is possible to underestimate the actual steady-state permeation rate if prediction models are used.

The study of the interaction between volatile organic compounds and packaging materials is complex. Initial studies have used simple systems in an effort to characterize packaging materials relative to one another. This may give useful information when only one compound is of concern. Further research is needed to develop more sophisticated methods for the measurement of multiple permeants. It must also be understood that some of the interactions which occur can cause changes in the permeability of the material being tested, resulting in the necessity to carry the tests out until steady has been reached. Once the interactions between a given polymer and permeant are understood, prediction testing can be used to shorten the length of the test.

Literature Cited

1. Annual Book of ASTM Standards; Standard D1434; American Society for Testing and Materials: West Conshocken, PA, 1994; Vol. 15.09, pp. 204 - 211.
2. Annual Book of ASTM Standards; Standard D3985; American Society for Testing and Materials: West Conshocken, PA, 1994; Vol. 15.09, pp. 542 - 547.
3. Annual Book of ASTM Standards; Standard F1249; American Society for Testing and Materials: West Conshocken, PA, 1994; Vol. 15.09, pp. 1051 – 1055.
4. Hirose, K.; Harte, B.R.; Giacin, J.R.; Miltz, J.; Stine, C. In *Food and Package Interactions*; Hotchkiss, J. H., Ed.; ACS Symposium Series 365; American Chemical Society: Washington, D.C., 1988; pp. 28 – 41.
5. Moshonas, M.G.; Shaw, P.E. *J.Fd. Sci.* 1989, 54, pp. 82-85.
6. Sadler, G.D.; Braddock, R.J. *J. Fd Sci.* 1991, 56, pp. 35 - 37.
7. Nielsen, T.J.; Jagerstad, I.M,; Oste, R.E.; Wesslen, B.O. *J. Fd Sci.* 1992, 57, pp.490 – 492.
8. Charara, Z.N.; Williams, J.W.; Schmidt, R.H.; Marshall, M.R. *J. Fd. Sci.* 1992, 57, pp. 963 – 966, 972.
9. Konczal, J.B.; Harte, B.R.; Hoojjat, P.; Giacin, J.R. *J. Fd. Sci.* 1992, 57, pp. 967- 972.
10. Risch, S.J. *Fd. Tech.* 1988, 42, pp. 95 – 103.
11. Hernandez, R.J.; Giacin, J.R.; Baner, L.A. *J. Plastic Film & Sheeting,* 1986, 2, pp. 187 – 211.
12. Arora, D.K.; Hansen, A.P.; Armagost, M.S. In *Food and Package Interactions II*; Risch, S.J. and Hotchkiss, J. H., Eds.; ACS Symposium Series 473; American Chemical Society: Washington, D.C., 1991; pp. 203 – 211.
13. Pasternak, R.A.; Schimscheimer, J.F.; Heller, J. *J. Pol. Sci.*1970, 8, pp. 467 – 479.
14. Annual Book of ASTM Standards; Standard F1769; American Society for Testing and Materials: West Conshocken, PA, 1998; Vol. 15.09.

Author Index

Subject Index

Highlights from ACS Books

Desk Reference of Functional Polymers: Syntheses and Applications
Reza Arshady, Editor
832 pages, clothbound, ISBN 0–8412–3469–8

Chemical Engineering for Chemists
Richard G. Griskey
352 pages, clothbound, ISBN 0–8412–2215–0

Controlled Drug Delivery: Challenges and Strategies
Kinam Park, Editor
720 pages, clothbound, ISBN 0–8412–3470–1

Chemistry Today and Tomorrow: The Central, Useful, and Creative Science
Ronald Breslow
144 pages, paperbound, ISBN 0–8412–3460–4

A Practical Guide to Combinatorial Chemistry
Anthony W. Czarnik and Sheila H. DeWitt
462 pages, clothbound, ISBN 0–8412–3485–X

Chiral Separations: Applications and Technology
Satinder Ahuja, Editor
368 pages, clothbound, ISBN 0–8412–3407–8

Molecular Diversity and Combinatorial Chemistry: Libraries and Drug Discovery
Irwin M. Chaiken and Kim D. Janda, Editors
336 pages, clothbound, ISBN 0–8412–3450–7

A Lifetime of Synergy with Theory and Experiment
Andrew Streitwieser, Jr.
320 pages, clothbound, ISBN 0–8412–1836–6

Photochemistry and Radiation Chemistry

James F. Wishart, Editor
448 pages, clothbound, ISBN 0–8412–3499–X

For further information contact:
Order Department
Oxford University Press
2001 Evans Road
Cary, NC 27513
Phone: 1-800-445-9714 or 919-677-0977
Fax: 919-677-1303